Doug + Jane Brachu

November 1977

THE BIG WOODS

Logging and Lumbering—
from bull teams to helicopters—
in the Pacific Northwest

Books by Ellis Lucia:

THE BIG WOODS

OWYHEE TRAILS

MR. FOOTBALL

THIS LAND AROUND US

THE SAGA OF BEN HOLLADAY

KLONDIKE KATE

TOUGH MEN, TOUGH COUNTRY

HEAD RIG

CONSCIENCE OF A CITY

THE BIG BLOW

DON'T CALL IT OR-E-GAWN

WILD WATER

SEA WALL

Included in anthologies:

TRAILS OF THE IRON HORSE

LEGENDS AND TALES OF THE OLD WEST

BITS OF SILVER

THE BIG WOODS

Logging and Lumbering–
from bull teams to helicopters–
in the Pacific Northwest

by ELLIS LUCIA

DOUBLEDAY & COMPANY, INC.
GARDEN CITY, NEW YORK
1975

Library of Congress Cataloging in Publication Data

Lucia, Ellis.
    The big woods.

    Bibliography: p. 212
    Includes index.
        1. Lumbering—Northwest, Pacific—History.   I. Title.
SD538.2.N75L82   338.1'7'498209795
ISBN 0-385-02461-4
Library of Congress Catalog Card Number 74–12699

For
HUGH McGILVRA

publisher, editor, friend . . .

who pointed at the Tillamook Burn,
the log trucks streaming through town,
and the tall timber, and said,
"That should be your 'beat' . . ."

Acknowledgments

Many people living in the Big Woods of the Pacific Northwest, in various walks of life and over a lengthy period of time, contributed to this book. It is impossible to list them all, and I can therefore only generally express my gratitude for their co-operation, interest, support, and encouragement.

Some, however, deserve special mention for taking the time and trouble to ferret out things of particular interest, consent to lengthy interviews, or for coming forth with ideas and material that enriched the over-all project.

I am particularly grateful to Alfred D. "Cap" Collier of Klamath Falls, Oregon, and Mrs. Nelson Reed for giving permission to use the wonderful *Ode to a Logger's Boots,* which sets the scene for this book. As an aside, Mr. Collier is founder of an extensive logging museum of huge old-time equipment in a state park north of his city, a marvelous place for seeing how it was.

From this point, it is difficult to know how to proceed. Early in the game, a longtime friend, Phil Dana, and his good wife, Jean, who help Carwin A. Woolley run the huge Pacific Logging Congress, gave me valuable "leads" about that side of the industry and its colorful history. Mr. Woolley also made available the PLC's extensive files, including early copies of the annual *Loggers Handbook,* and was always willing to answer my questions. Merlin Blais of Western Wood Products was another who exercised great patience in the question department, or who came up with the right name where I could get the answer. Many, many others contributed in a like manner.

Suzanne Richards Kalapus, digging in the library of her newspaper, the *Oregon Journal,* called my attention to the writings of Mary L. Roberts on the Tillamook fire, which led to a new look at that tragedy. Larry Skoog, field adviser for the Oregon Historical Society, had some valuable suggestions, too, about the fire and also special material concerning the Wobblies. Barbara Elkins, Society librarian, was most helpful as always, as were others of that outstanding institution. I am grateful to Hugh and Paul McGilvra and the staff of the *Washington County News-Times* for allowing me to clutter the office while digging into back files. A special word of "thanks" needs to go to William Lyda and his wife, Myra, for telling his version of how the Tillamook fire began; and to others who took their time to reminisce about the holocaust, among them Walter Vandervelden,

Arthur Reeher, Mrs. Max Reeher, and Clifford R. Johnston of Forest Grove, Oregon, and A. C. Johannesen of King City, Oregon.

An old friend, J. Edward Schroeder, the Oregon state forester, furnished much material concerning present-day forestry in his state, and also reflected upon the early years of the Tillamook Burn reforestation program. Frank Sargent, deputy state forester, clarified other points of puzzlement. Bill Hoskins of the Northwest District spent a day guiding me through the new Tillamook forest and telling about the problems of modern forestry. Jim Fisher of the Oregon Forestry Department was also most co-operative. And Warren Warfield, staff forester, forest land management division, Washington State Department of Natural Resources, contributed much valuable material. Merle F. Pugh of the U. S. Forest Service office in Portland was helpful in checking some fine points and introduced me to deputy regional forester Robert H. Torheim, who spent a long while telling of forestry problems in this computerized age.

Lamar Newkirk and Steve McNeil of Georgia-Pacific were most helpful in guiding me through the maze of the modern timber industry and industrial forestry, and in explaining their company's new shotgun planting tool. Robert Lindsay and Jack Brown were likewise of much assistance at Crown Zellerbach, and Mr. Brown came up with some colorful material on the early paper industry in the Pacific Northwest. Guy B. Pope detailed the modernization of Pope & Talbot, while Mrs. Pope furnished rich material concerning the company's early history and the refurbishing of old Port Gamble.

Larry Williams, executive director of the Oregon Environmental Council, explained the viewpoint of ecologists and/or preservationists toward public and private forestry, logging, clear-cutting, etc., in the Big Woods. Courtland Matthews, retired editor of *Chain Saw Age,* generously turned over his file on the history of the power saw to me, which was most helpful in filling out the details of that subject. Albert Wiesendanger, longtime executive secretary of Keep Oregon Green, supplied much material and spent several hours detailing his long career. Ed Loners of Keep Washington Green was likewise most co-operative. Wes Lematta of Columbia-Construction Helicopters gave me a detailed account of the problems of helilogging, while Ernest L. Kolbe, manager, furnished information about the new Western Forestry Center. Jack Pement of the *Oregon Journal* deserves a word of thanks for permitting my use of his material on Dorothy Ann Hobson.

The staff of the Multnomah County Library was helpful, while the Oregon State University library supplied "leads" about power saws. Others who contributed include: John E. Benneth, American Forest Institute; Irving Luiten, the Weyerhaeuser Company; Will Rusch, McCullough Corp.; David Koonce, Omark Industries; W. W. Hagenstein, Industrial Forestry Association; Samuel R. Donaldson, Boise Cascade Corporation; W. S. Looney, Simpson Timber Company; and Don Woodman of Portland.

Last, a tip of the logger's red hat should go to my good wife, Elsie, who brought home reports about the timber industry, called my attention to articles and anecdotes, and in addition to her own busy job, did the final typing of the manuscript. This latter, of course, also gives her license to kibitz and make valid criticisms, and ensures her a first copy of the published book.

ELLIS LUCIA
Portland, Oregon

Foreword

Until comparatively recent years, and before such words as "smog" and "ecology" were common to the language, a regular feature of the Pacific Northwest's summer and autumn scene was the smoky haze that hung over much of the region. At times, this pall grew very heavy indeed, when a huge forest fire was on the run or when there was extensive slash-burning—slash being the rubble left from the thousands of logging operations working the vast forests of this timbered Northwest, which early settlers called the Green Desert. Another contributor to the smoky summers was the hundreds of wigwam-shaped burners that had been ingeniously designed to take care of the wood waste that almost buried the sawmills and were the landmark of nearly every lumber plant, large and small, in the territory.

Few natives outwardly complained, even when the smoke settled low over the villages and towns, sending soot down upon freshly laundered clothing or when live sparks fired rooftops and vacant lots. This was a way of life; that pall was part of the brawling, two-fisted industry that dominated the region and kept the dollars rolling. The only outcries were against the recklessness that caused forest fires, for thousands of rich acres were going up in flames, which not only would shorten the years of lumbering in these last great timber stands in the United States, but was also wanton destruction of a resource that could serve for generations. But it seemed little could be done about the fires, which came from lightning and public carelessness as well as from logging.

The smoky haze belonged to a freewheeling style of living, along with the rumble of log trucks through the little towns, the long pole trains snorting up the grades, and the scream and bang of sawmills in the night. You couldn't get away from lumbering, especially if living anywhere west of the Cascade Mountains. Nearly every town and crossroads had its sawmill, and many of today's big centers, most notably Seattle, accommodated sawmills from the very first. The whining mills, many of them giving the appearance that they would collapse in the slightest wind, were scattered everywhere over the land, and I recall one located right on the broad beach of the southern Oregon coast, a fact that would surely set beach-lovers to howling today. Farmers often combined their agricultural pursuits with the harvesting of timber, perhaps in a small gyppo logging operation or running a little mill for rough-cut.

Everyone lived with timber, for the Big Woods furnished the very sustenance of life itself. You cooked and heated your home with wood, or its main waste product, sawdust; and I well remember the great mountains of sawdust that provided fuel for the steam electric generator plants producing city power. And you knew when autumn was on the way, hearing the zing-zing of the small portable gas-driven saws cutting up the huge piles of slab in the neighborhoods for winter fuel. At times these towering stacks of slab in the parkings made tunnels of the narrow streets, especially in Portland's northwest side, which gained the name of Slabtown.

The cutting of trees and the manufacture of lumber dominated every aspect of society, from the economic, political, and recreational activities of the people who worked in or near the industry to the manner in which their offspring were educated. The lumberman was king; and the logger, in his red hat and high-water blue jeans, was the crown prince. Their rule affected everything, from the way the towns were laid out to a minimum of highway safety laws for the logging trucks. The trucks not only broke down the roads but were also continually dropping their logs, often killing and maiming luckless motorists who happened to be passing at the precise moment when a load let go, yet the operators were angered at any thought of restrictions on how large a load they could carry or how many chains should be used to hold the logs in place. They were an independent breed who packed a wallop with all the lawmakers.

The little town of Forest Grove, population twenty-three hundred when I spent my salad years there, could hardly be called a typical logging-lumbering community, although this was certainly a major part of the local scene. Forest Grove was also the setting for a rich agricultural area and a small university, which had as a major claim to fame a bruising football team comprised of tough young loggers, and loggers' sons, who got into shape for the fall gridiron schedule by working in the woods. It was indeed a strange conglomeration along the downtown streets of loggers, farmers, shopkeepers, college students, and professors. Yet even here, the industry was a dominant force. The logging trucks growled daily along the two main streets, swinging wide for a right-angle turn at the main intersection and sometimes clobbering a car parked too near the corner (or maybe because the log was extra long), and occasionally dropping logs onto a parked car, turning it in seconds into a candidate for the scrap heap. Many of those logs were an ugly coal-black, salvage from the sprawling Tillamook Burn a few miles to the west; only seven years before my arrival the region had sustained the greatest forest fire of them all, one that altered its destiny for generations to come. Otherwise, Forest Grove might have been more deeply involved in lumbering than it was.

Sawmills rimmed the town and were scattered over the small valleys and hills, and there were numerous logging operations, providing a substantial payroll. The sawmills were a constant concern of the volunteer fire department, for the flimsy plants had a reputation for catching fire, from their burners if nothing else. Since the holocaust of 1933, the volunteers had been even more fire-conscious; the year after the blowup, they purchased with their own money—the city funds were depleted—an old-fashioned steam pumper being retired by the city of Portland, so that they might better handle any mill fire in their vicinity.

You went to sleep to the song of the sawmills and the distant rumble of logging trains moving in darkness toward the log dumps along the Willamette and Columbia rivers. The scream of the head rig, the twang of the resaw, the slam and clatter of boards along the green chain, the shouts of the men, the glow of the wigwam burners—these were healthy signs of industry in whatever locality you were living. Nevertheless, things in Forest Grove didn't always go as serenely as they did in company towns and mill towns devoted entirely to this raw-boned industry. There were complaints against the fallout from a nearby mill, which had a poor burner that dirtied laundry, windows, houses, and yards of residential districts, and also about the noise pollution (the term hadn't been created yet) caused by "that darned mill." Heavily loaded logging trucks were forced to round a sharp curve near an elementary school, and at the same time needed to shift gears to climb a slight grade. The resulting snorting and blowing constantly interrupted classes, and the principal even gathered some impressive statistics to show how much was being lost in teaching time, at taxpayers' expense. But this wasn't a problem unique to The Grove. My wife recalls that in her own school days in another small community, the racket of the sawmill just across the way at times made oral classwork impossible.

Much more serious at Forest Grove's grade school was a remarkable condition that sent logging truckloads flying hell-bent onto the school grounds every so often when rounding that whipping curve. Luckily, no youngsters were ever around when this happened. The public got up in arms about it, as well it should, so the curve was realigned until the highway could eventually be moved away from the school, and the haulers pledged to chain their logs better and take the curve more gracefully.

These slight community matters may seem trivial today, but they were nevertheless straws in the wind, signifying what was ahead for the Big Woods. The forests and the industry were on the verge of the greatest changes and pressures since John Dolbeer's steam donkey knocked out the bull teams. America's first industry, which moved across the continent with the early settlers, was thought by many to be beyond taming; even the wildest dreamers could not imagine the advent of full mechanization and the use of advanced technology, including computers, for everything from timber cruising to sawmilling and the rapid regrowing of the forests. It's still tough, dangerous work, but the old-time bull-o'-the-woods would hardly recognize his hard-hat counterpart of today, who labors with levers in a snug cab, fells, bucks, and loads logs in less time than it took a tail down man to spit, travels to and from his job in his own car, and lives with his family in a comfortable house where he puts his feet up on the coffee table like everybody else to watch the Saturday ball games.

The woods, the sawmills, the great processing plants, the tree farms and nurseries, the attitude toward what the public calls ecology and a healthy environment—all are different from the old ways, although the basic things are still there. But most of the small independent mills are gone, and the traditional gyppo logger is fading away, for it takes a fortune today in modern equipment to get into the logging and lumbering business. In the intense ecology movement of the 1970s—and nowhere is it more aggressive than in the Pacific Northwest—the requirements for cutting timber are becoming more rigid all the time. The logger of yesteryear dreamed of a skyhook for his rigging. Now logging is done with helicopters and balloons, a result of concern for the environment and the tearing up of the land.

We don't have great forest fires anymore, brought about by stricter laws on logging in dangerous weather, spark arresters and other safety devices, a change of attitude by the public toward its own responsibility in the forests, far more and better fire-fighting equipment, air patrols and smoke jumpers, and the ability to hit fires hard at their first outbreak. This changed the picture as to interest in replanting the forests, so that through the tree farm movement and scientific forestry, there is small reason for not having vast forested areas for generations to come.

There are no longer huge quantities of waste. Forest laboratories have developed a wide array of products, utilizing wood chips and fiber for everything from hardboard paneling to many varieties of packaging. Lumber in its simplest form is now only part of the picture. Hog butchers have long boasted of using everything but the squeal; lumbermen now brag about full utilization of everything, including the bark. Big choppers are returning slashings to the soil of denuded forest lands, on which new generations of trees will nourish and develop. Some timber companies boast of a new supertree that will bring about better forests faster, under the slogan coined long ago that "timber is a crop." All stops are out; even aerial photography is used in the planning, growing, and harvesting of trees. Still, in the environment-minded 1970s, there is much bitterness, for the logger and lumberman remain the villains in the public mind, the conservationist the good guy, and the forester caught somewhere in between. The extreme conservationist will advocate locking everything up, never to cut another tree, and then go home to relax in his woodworking shop. At the other end, a wild-eyed lumberman would cut trees "as I damned well please," and of course his attitude, linked to a rowdy past, makes him his own worst enemy. The public meanwhile feels it has a big stake in the forests and can become pretty irate, usually at the lumberman. So the debate rages on, and there is much emotionalism as this book is completed.

The other day I was chatting with a veteran logger, Clifford R. Johnston, who has been in the woods for over fifty years, and has done just about everything. He started washing dishes in a logging camp at age twelve and then graduated to a whistle punk. His father logged with bull teams until he was seventy. In his middle sixties, Johnston was still going strong. Yet he met me at the door dressed in sports shirt and neatly pressed slacks, his Oxford-style shoes shined to a spit and polish. His home was comfortable, neat, and modern, and on display were framed photographs of his children and grandchildren. He lives like any other average citizen of his town and might easily pass for a businessman. The only indications of his trade were his knowledge of the subject and occasional resorting to a logger's unique verbiage.

"I've seen it all," remarked Johnston, summing it up, as loggers do, in a very few words. "There have sure been some changes."

And that, dear reader, is what this book is all about. . . .

ELLIS LUCIA
Portland, Oregon

Contents

Oh, Stranger, ponder well what breed of men were
these cruisers, fallers, skinners,—ox, horse and 'cat,
chokersetters and the rest who used these tools.
No summer's searing dust could parch their souls,
Nor bitter breath of winter chill their hearts.
'Twas never said "They worked for pay alone," tho it
was good and always freely spent. Tough jobs to
lick they welcomed with each day. "We'll bury that
old mill in logs," their boast. Such men as they
have made this country great, beyond the grasp of
smaller, meaner men. Pray God, Oh Stranger, others
yet be born worthy as they to wear a logger's boots.

Ode to a Logger's Boots
By NELSON REED
Collier State Park
Logging Museum
Klamath Falls, Oregon

PART ONE

UNTAMED FELL AND BUCK

Chapter I

Whar's Yer Gold, Johnny?

"On the verge of the glowing horizon, I discovered the majestic dome-like crowns of Big Trees towering high over all, singly and in close grove congregations. There is something wonderfully attractive in this king tree, even when beheld from afar, that draws us to it with indescribable enthusiasm; its superior height and massive smoothly rounded outlines proclaiming its character in any company; and when one of the oldest attains full stature on some commanding ridge it seems the very god of the woods. . . . I made my bed and supper and lay on my back looking up to the stars through pillared arches finer far than the pious heart of man, telling its love, ever reared. Then I took a walk up the meadow to see the trees in the pale light. They seemed more marvelously massive and tall than by day, heaving their colossal heads into the sky, among the stars, some of which appeared to be sparkling on their branches like flowers."

JOHN MUIR
The Wilderness World of John Muir

When the cry of "Gold! It's gold, by Gawd!" echoed from California's Sierra foothills, many a bored farmer in the Oregon Country far to the north unhitched his team, left his plow in midfield, packed his gear, and headed south, abandoning ranch, wife, and youngins to hunt for treasure in the creeks.

Among them was John Porter, who had settled obscurely on the West Tualatin Plains, an offshoot of Oregon's great and fertile Willamette Valley. Like all the others, Porter dreamed of get-rich-quick results in the Mother Lode where he'd heard the rainbow's end was anchored and you could pick up nuggets big as your fist right from the surface of the ground. Then the bubble burst; somehow reality didn't carry the vivid sparkle of the dream. Porter discovered, as did thousands of others, that groveling for gold had its shortcomings. The hours were long, the disappointments many, the pickings slim, and more than likely you wound up with an aching back.

Porter, who had the characteristics of a poet, began wandering through the foothills and mountains of the warm Sierras, halfheartedly looking for nuggets but also attracted by the region's beauty and solitude. One day Johnny stumbled upon an awesome sight—a tree so straight and tall it seemed to support the sky. He couldn't believe it. He'd never seen anything so grand, of such astounding girth and height, even in the Oregon Country. Perhaps it was a freak of nature. Then he found more of them, the lofty *Sequoia gigantea* of the Sierra Nevadas. He guessed rightly that they presently held a thousand years of life, the largest and oldest living things on earth.

When Johnny returned home, two bulging sacks were slung across his saddle. Family and friends rushed out to see Johnny's treasure. Porter dumped the contents of a sack on the ground. Jaws dropped; there was nothing but some strange cones.

"You must be plumb crazy," a neighbor shouted. "Whar's all yer gold, Johnny?"

Porter stammered. Standing amid the cones, he appeared ridiculous, even to himself. He couldn't explain about the glorious big trees he'd seen in California. When he tried, holding cones in each hand, the settlers only laughed. How do you talk about big trees in a land of tall timber, much of it of such size to dwarf anything in the world, and where trees are considered a nuisance, save for building cabins, barns, and hogpens?

Nevertheless, Porter planted the cone seed on his place and gave the young shoots tender, loving care. His nursery was the nation's first tree farm. When the seedlings were old enough, he set out a grand avenue of trees leading to his house. He distributed other seedlings in the fashion of Johnny Appleseed to people who promised to care for them. Some were placed on the grounds of the Washington County Courthouse, more given to the Catholic church of the nearby Dutch settlement of Verboort, and still others to the town of Forest Grove, whose very name was inspired by its surroundings.

More than a century later, John Porter's tiny seedlings have become lofty giants, visible for many miles and standing distinctively above all else. Sequoias are scattered throughout the Willamette Valley, from Porter's stock or brought back by others from the Mother Lode, for Johnny Porter had started something. The transplants thrived on the rich, rain-washed plains of this Northwest land, which was meant for growing trees. A mile north of Forest Grove stands the finest display of all: a great cathedral hall of nearly forty sequoias, a hundred yards long on both sides of a thin road and of rare and indescribable beauty, unlike anything else found anywhere, a memorable sight to visitors who pause to reflect on men like John Porter.

Yet the legacy of Porter's sequoia giants doesn't end there. Ravaging storms and explosive lightning have taken their toll, although most of the giants have managed to escape the destruction of twentieth-century "progress." But in 1948, ironically just a century to the very year that California gold was discovered, a homeowner in Forest Grove, on the edge of the great Tillamook Burn, ordered two of the historic giants in the front yard of her 1883 pioneer frame house to be cut down.

Mrs. D. M. Macleod declared grimly that she had put up with the sisters long enough. The trees, nearly a century old, were still infants by their genus, although about 150 feet tall and spreading in all directions.

"They just keep growing and growing," complained Mrs. Macleod.

The twins were pushing against the house, which had been moved once to accommodate them. The lady refused to have it done again. Their root systems were breaking up the foundation and cracking the retaining wall.

Neighbors and Oregonians everywhere were horrified by the woman's decision, for the sequoias were something held in sacred trust for generations to come. The chances are that in the ecology-minded 1970s, her decision would have brought organized protests and sign-carrying marches, for there are a great many people who believe it is a sin to cut any tree, let alone one with a rich historical heritage.

"I spent my girlhood under those trees," said Mrs. Macleod in her own defense. "They have as great a sentimental attachment for me as anyone."

Four tree surgeons were hired from Portland for the execution. Felling the giants proved to be quite a ticklish logging operation, not only from their massive size but because this was a residential neighborhood, with houses nearby. The trees couldn't just be "dropped," as in the woods. The majestic giants, which might have lived several thousand years, were stripped naked of their graceful branches during many sad days of preliminary work. A high climber topped 37 feet from the taller of the pair, which gently touched the clouds 148 feet above the ground, so that in the final fall it would miss surrounding wires. The death throes were accompanied by moans from the public and photographs and stories on the front pages of the newspapers, spread afar by the wire services.

A crowd gathered to witness the sorry execution. Using a giant chain saw, the lumberjacks began cutting into the 82-inch butt of the taller sequoia. The noisy saw sputtered carbon monoxide and stopped several times. Then the lofty giant wavered and crashed to the ground, rattling houses and windows for blocks around as if there'd been an earthquake. A great cloud of dust drifted off; there was a huge gash in the ground of the vacant lot next door. Soon the sister tree was lying nearby, and people wandered away, feeling a sense of loss they didn't quite understand.

There were many letters and telephone calls. Editors wrote lengthy obituaries bemoaning the deaths of the giants. A lumberjack named Joe B. Cox, who had invented a new type of saw chain, came out to test it on the logs.

Part of one log was turned ironically into wooden money souvenirs for an upcoming centennial celebration of nearby Pacific University. Then for almost a year the massive logs, cut into eight-foot lengths at the time they were toppled, lay in the yard where they had fallen. Mrs. Macleod had much trouble locating a sawmill that could handle the giant logs weighing three to seven tons and estimated to yield some 10,000 board feet of good lumber. At long last the following summer the sequoias were trucked, with struggle, to the Olson Mill in Scoggins Valley, south of town, amid more headlines. The entire operation cost the homeowner a sizable piece of change, but Mrs. Macleod, who by now was regretting her decision, was glad to be rid of the remains. In the months following their demise, she had suffered much public abuse. Cruel things were said verbally and in letters, and on reflection, she confessed to me that had she realized what she knew now, she doubted that she would have had the courage to destroy the trees.

Yet in that single episode, on a quiet residential block of a small Oregon town, was played the entire drama of logging and lumbering. John Porter, the pioneer conservationist, attempted with his strange nuggets of green gold to improve upon nature in his homeland, where big trees are commonplace, yet the love for them is as intense as anywhere. Mrs. Macleod in another age dared destroy part of a heritage in a locale where Porter's legacy was real and alive. Although the trees stood on her property, the public felt a keen sense of ownership and responsibility for the sisters.

"The thing to have done, if anything were to be done," editorialized *The Oregonian,* "would have been for the state to purchase the property and endow the sequoia sisters. Think of them as they should have been, a dozen centuries hence!"

It was the cry of the ages, "Woodsman! Spare that tree" as the gamut of lumberman vs. conservationist was run in the felling of the sequoias. The lady suffered the kind of abuse all too familiar to loggers and lumbermen in their everyday work. Lumber is a basic need, yet people dislike seeing trees cut down and the destruction of forests, although not entirely without sound reasons. But there is also much emotionalism and breast-beating, since trees are considered by many to be God's handiwork, as projected in Joyce Kilmer's celebrated poem. Therefore, destruction of the sequoia sisters symbolized something far broader in scope to this lumber-producing region, where many of Mrs. Macleod's neighbors were on logging and lumbering payrolls.

Ironically, the Macleod pioneer home still stands, having long outlasted the big trees that got in its way. And many of John Porter's majestics tower in splendor in the valley. Yet like most of the pioneers, Porter couldn't see the forest for the trees. He found the beauty but not the gold that was all around him in his homeland, worth billions in rich veins that covered the hills to the highest peaks. Only the color was green, not yellow, and it would take a different breed of men to mine it.

Chapter II

The Great Green Desert

With respect to the appearance of the country, my expectations were fully realized, in its fertility and variety of aspect and of soil. The greater part, as far as
the eye could reach, was covered with Pines of various species.

David Douglas

In the Pacific Northwest, the vast forests are a way of life. East of the Cascade
Mountains, the lodgepole and Ponderosa pines march in stately legions up the
steep slopes to the snowline, and stretch in great platoons along the ridges and
rolling foothills, and over the high plains into Idaho and Montana. In the pine
country, the forest floors are clean and free of undergrowth, and the big trees
are spaced apart, with many open clearings and yellow shafts of light hurtling
down onto the dry, soft carpets of needles, and with the rich, pungent fragrance
of the trees a wondrous thing during the dull, dry heat of summer.

But west of the Cascades is another kind of world. From Alaska and British
Columbia to the redwood country of northern California lies the Big Woods.
This is Douglas fir country, between the backbone mountains and the thunder-
ing coast, the trees growing so heavy in numbers that they crowd and elbow one
another, and their branches become entangled, so that many times, two or more
grow as one. Early-day trappers and settlers cursed and sweated as they rammed
their way among the massive plants and heavy brush of salal, salmonberry, fern,
and huckleberry, where even Indians and wild animals found the going nearly
impregnable, dark, and foreboding. The Indians set fire to the forests just to open
them up for game and hunting. Except in such places, the soggy forest floor containing the compost of centuries hadn't seen the sunshine for hundreds of years.

"Thick as hair on a dog's back . . . reaching to God's elbow" was the way
early settlers described the great woods. William Clark of the Lewis and Clark
Expedition noted how the heavy growth hampered their hunters. Trappers and
settlers preferred traveling on the rivers, since cutting across country through the
forests was a next-to-impossible undertaking. Few found the mighty timber

stands to their liking, save for a man like David Douglas, the Scottish botanist who was drawn several times to the Pacific Slope between 1825 and 1833 to study forests of some half hundred varieties of trees and especially the fine tree that today bears his name, describing it as "one of the most striking and truly graceful objects of Nature." It was originally known as the Oregon Pine, and because of the climate it grew to a monstrous size, along with the spruce, cedar, and hemlock.

Yet to the early comers—farmers and developers of trade centers—the trees were only weeds, a plagued nuisance that had to be cleared out before a fellow could plant a crop or build a town. Only rarely did a pioneer go to living in a giant hollow stump, although certainly some loners did. Early villages were built over and around the many stumps, for it was long and tedious work to grub the stumps or blast them out with gunpowder. The metropolis of Portland, Oregon, was originally known as Stump Town. Its multitude of stumps was whitewashed for visibility at night to wagons and celebrants staggering woozily home from Bill Johnson's riverside cabin, where he manufactured a fearsome concoction called Blue Ruin. Residents paid their taxes by grubbing stumps for the city, and those running amuck of early pioneer law were also sentenced to stump duty.

Everywhere the pioneers looked was that damnable green. It dominated all else, and while at first they'd been happy to see it after the dreary spacelessness of the arid plains, they began wondering now whether all this was a camouflage of fool's gold. The forests were utterly worthless, save for the fact that trees were handy for constructing buildings and for fuel. Most pioneers saw the world in terms of a plow. The broad plains of the Willamette Valley and all the smaller valleys were quickly staked out, for there the trees were in scattered woodlots, with much open space. But if a man couldn't get such acres—and many latecomers couldn't—he tried grubbing the stumps from quarter sections for plowing along the foothills. The damned seedlings popped right up behind the plow, so that it was a constant battle holding back the forests. Much of the land was found useless for growing anything but trees anyway, for it was rugged and rocky, and often standing on end. The people crowded into the open valleys and low hills so that today the bulk of the population of Oregon and Washington is bunched into a tenth of the area, with more than half the land still devoted to timber.

The pioneers stayed clear of the deep forests with their blackness, damp cold, unknown dangers, and utter, silent desolation. The Big Woods made a useless wasteland, save for hunting, and on a par with the dry, open deserts, except that the color and the climate were all wrong. The color brown and the waterless lands were held synonymous with the desert. Yet the Big Woods was green and as damp as a swamp. Green and water were normally associated with agricultural richness and lush productivity. The simple farmers were confused by this upside-down way of things as they looked at all those trees, all that land going to waste. They called it the Green Desert.

Only ships' captains held admiration for the many tall, straight conifers that appeared as sentinels along the deep-water bays and inlets, branchless from the ground for a hundred feet and more. In each the skippers saw a good ship's mast. John Meares, the early fur trader and ship captain, in 1788 felled a deckload of spars at Vancouver Island in what was British Columbia's first logging operation. Then Meares lost the whole shebang, forced to jettison the load during a violent mid-Pacific storm. While charting Puget Sound in 1792, George Vancouver replaced a mast on the *Discovery*. In the age of the windjammers, Northwest timber was much to the liking of skippers and owners for keeping the tall masters in first-class shape. But beyond this there was little vision.

Dr. John McLoughlin, the shaggy chief factor of the Hudson's Bay Company post of Fort Vancouver along the Columbia River, did indeed see the potential, for McLoughlin realized the full value of world trade. A few miles upstream from the post, McLoughlin's men erected the Northwest's first sawmill—a flimsy, water-powered plant of machinery shipped from far-off England. After frightening the local Indians with its strange convulsions, the little mill began producing 3,000 feet of lumber per day. Within two years, McLoughlin was sending lumber upstream and also his first export shipment of "Oregon Pine," as he called it, to the Hawaiian Islands at $80 per thousand board feet. By 1833 Fort Vancouver had a foreign lumber outlet to China, bringing much-needed revenue to the Hudson's Bay post on the declining fur market and also launching what grew into the great Pacific trade.

Other small sawmills, taking McLoughlin's cue, were springing up here and there on the edge of the timber and down in the Willamette Valley. They were primarily for local consumption, as there was no profit cutting for a man's neighbor who couldn't pay a price and was up to his eyeballs in trees of his own. In the midst of plenty, lumber was scarce and often hand-cut; what few small sawmills existed were crude affairs, slow up-and-down muleys that could turn out only a million board feet a year. Trapper Ewing Young retired from the Rocky Mountains to set up the first one in Oregon's Willamette Valley. Members of Jason Lee's mission built a sawmill at Salem and then got a second going at the Willamette Falls. Men of enterprise and muscle were slicing logs throughout the valley and on the coast, all filling the local needs of population growth as more pioneers pushed across the plains.

Among those arriving with hopes and dreams of a new life in the Promised Land was a Louisiana-born black and his white wife. George Washington Bush had seen the Oregon Country once before when on the payroll of the Hudson's Bay Company as a fur trade *voyageur*. He longed to settle there. Bush was a free black of considerable wealth as heir to the fortunes of an elderly couple under the care of his parents in their declining years. He had a good education in Pennsylvania, served with Andrew Jackson at New Orleans, and later was successful in farming and cattle buying in Missouri.

Then Missouri passed a law making it illegal for free blacks to live in the state. Bush remembered Oregon, with its promises of living in freedom and dignity for his family. Part of the Northwest would likely become British, where slavery wouldn't be permitted. Bush and his family headed West with his close friend, Michael T. Simmons, a 30-year-old illiterate Army colonel who wanted to take his family of seven youngsters to Oregon. In 1844 the pair organized a wagon train, using Bush's money and Simmons' ability to command. Bush brought along farm implements, herds of stock, nursery cuttings, and seed. He bolstered poor members of the train by loaning teams, provisions, and money, and with his spirit of optimism. Bush was considered to be "the wealthiest man to migrate to the Oregon Territory" to that time.

Then came the shock. Bush's arrival was an upsetting thing to the settlers already established in the Willamette Valley, many of them from Missouri. The new provisional government passed a sweeping law barring any person of Negroid blood to settle upon or own land. No black slaves could be imported, and all blacks and mulattoes were banned under penalty of arrest and flogging "once every six months until he or she shall quit the territory." Members of the Bush-Simmons train were angered by the anti-black law and stood firmly with the black man who had befriended them. Dr. McLoughlin was sympathetic, for he knew personally the miseries of race prejudice, as his first wife was a Chippewa Indian who bore him two sons. McLoughlin suggested that Bush locate deep to the north of the Columbia River, where authorities of the provisional government were less likely to bother him. Meanwhile, the Missourians were pressing hard for the complete removal of Bush and his family from the territory. The Bush-Simmons party headed north in what was the first break in Dr. McLoughlin's policy to "hold the line" at the Columbia River and keep the Americans in the South.

At the foot of Tumwater Falls in the summer of 1847, Bush and Simmons built the first sawmill on Puget Sound, the forerunner of hundreds in what would become a worldwide trade. Discarded machinery from Hudson's Bay, purchased with Bush's money, was utilized for the mill, which was purely a local venture. Bush and Simmons had trouble getting it going. Floodwater wiped out the first mill and pushed a second so out of kilter that it had to be torn down. Finally there was a workable plant just in time for the big Gold Rush in California. Small vessels could ride the high tide to this place, called New Market, and in the current demand for lumber, the brig *Orbit* called at Tumwater with supplies for a small settlers' store and hauled away the first export cargo of shingles and piling. The sawmill flourished, gaining a capacity of one hundred feet an hour. Later it was sold to Captain Clanrick Crosby of California, ancestor of an entertainer named Bing who first limbered up his vocal chords while working in the Northwest woods.

George Bush turned to farming, but his troubles with the Oregon government weren't ended. Simmons, who was now justice of the peace and was growing in influence, went to the territorial capital at Oregon City to plead his friend's case. Despite his illiteracy, Simmons was so eloquent and sincere that he managed to have a special act passed removing much of Bush's racial disability, although George was still unable to own property under the federal Donation Land Claim Act. Bush had also grown in stature in his own community. His farm at Bush Prairie was a hospitable way station on the route up Puget Sound. He helped many new settlers get started with his nursery stock and seed. During the battering winter of 1852, when famine was a reality, Bush was the only farmer on the Sound with a good supply of wheat. Seattle speculators offered him high prices, but Bush refused to capitalize on the misfortunes of others and instead apportioned his supply to his starving neighbors.

The people didn't forget. When the Washington Territory was created in 1853, Simmons and the others were quickly pounding on the legislative door, demanding a resolution to Congress giving Bush permanent possession to the land he'd developed over a decade. The memorial pointed out that "he had contributed much toward the settlement of the territory, the suffering and the needy never having applied to him in vain for succor and assistance. . . ." Congress passed a special act two years later, giving Bush full title to his land.

Still, it would take a special race of rugged men to discover value in all this timber, stretching from the slopes of the Cascades down to the sea. It was passing strange, too, since lumbering was America's first industry, beginning with Jamestown and in New England. The logger and the sawmill man followed the settler across the wilderness, helping to open up the country, for the trees had to be cut before the land could be tilled. The downed timber was ripe for cabins, corrals, bridges, stockades, and towns. The lumberjacks and the millwrights worked their way west to the mining camps, the cattle ranches, the stagecoach stations, and the river ports, and a bit later in the tremendous task of building railroads across the West. Wherever there were people, there was a sawmill somewhere close by, save on the treeless plains, where timber had to be hauled far distances. Therefore, it isn't at all amazing that when gold was discovered in the California foothills, exploding a madness that would bring the Pacific Northwest's Green Desert into proper focus, it was a sawmill man who made the discovery. John Marshall was building a sawmill for his boss, Captain John Sutter, when he picked up the first nuggets. He touched off a holocaust, but it was quite characteristic in changing the course of history. Many times the tough lumberjack and shaggy lumberman have turned the order of events, yet unlike the miner, the cattleman, and the gunslinger, they have never been romanticized in the Western saga. The reason is clear. The logger worked in isolation, far back in the brush, and often under climatic conditions too miserable to consider; besides, he was something evil, an uncivilized animal who felled trees and destroyed land. It was far easier to understand a rancher, a cowhand, or a miner than these timber beasts.

The ship that broke the news of the California gold strike added a footnote: Lumber was in great demand. The poorer sticks would bring $60 per 1,000, while good, clear stuff ran to $120 per 1,000, compared with $30 locally. Suddenly settlers of the Oregon Country found a fortune on their doorsteps. The Green Desert took on a new luster, with treasure hanging from every branch. Yet it wasn't that simple; the flimsy sawmills, numbering about thirty in all, couldn't handle the sudden need. Fortunes were slipping through the fingers of the poor farmers, for they had neither the equipment nor the know-how to turn the big trees into boards.

Now the lumbermen were arriving, and they weren't about to allow such a bonanza to stand idle of good use. Just ahead of the Gold Rush, Stephen Coffin induced a millwright, W. P. Adams from New Hampshire, to come to Portland, Oregon. Coffin was interested in developing Stump Town, where he had heavy land investments. The pair teamed up with Cyrus A. Reed, a schoolteacher and landscape artist who longed to become a sawmill operator. The plan was to build a steam sawmill with considerable capacity, the first in the region; and the three partners were able to cash in on some of the Gold Rush demand before the mill burned to the ground, a faulty characteristic of these flimsy plants well beyond the mid-twentieth century.

On Puget Sound, Henry Yesler announced that he would construct a steam sawmill to produce boards, dollars, commerce, and industry for the little village called Seattle. The town's founders, eager for anything that would expand their platted development, generously presented Yesler with a shovel-shaped timber tract of 320 acres, all heavily forested and extending from the water's edge back up the hill. The shape of the claim would provide Yesler with ready access to deep water and to the hillside, no matter what sprung up around him, which turned out to be a great deal. Yesler's idea for steam power made good sense, too, for one difficulty in mining this green gold was finding sufficient water power and stout machinery to bite through the big pitchy stuff. Yesler dashed to San Francisco to purchase machinery and before very long was off and running, snaking his logs down the great hill with oxen, over other timbers laid crosswise to the trail. This was the first skid road, along which in time sprang up a conglomeration of shacks and shanties of various enticements for the loggers. Yesler himself erected a massive cookhouse, which became a landmark as Seattle's nearest approach to a hotel, with a great circular saw hung outside as a dinner gong and for signaling starting and quitting time at the mill.

Came the dawn. Gradually, the astounding potential of this sprawling evergreenery emerged into clear vision, due primarily to the mounting demands in California. Millions could be made here by those who had the know-how and the stamina. Lumbermen from New England certainly had the know-how, drawn on several generations. But Maine was too far a haul, even at San Francisco's high prices, to serve as the woodbox of the West Coast. Andrew Pope and Frederic Talbot, members of Maine lumber families, were shaken by the great timbers they saw in San Francisco Bay in the network of piers extending far into the water. But ships' captains like grizzled Lafayette Balch, engaged in moving these huge sticks, weren't talking about where they came from, although Pope & Talbot had a hunch it was Puget Sound, where such timber (they were told) grew right to the water's edge and thus could be handled with reasonable ease.

The partners hoped to build much of San Francisco with "Oregon Pine." Captain William Talbot of the brig *Oriental* loaded his ship in Maine with lumber and made the long trip around Cape Horn to join his brother and partner on the Golden Shore. They also sent for their millwright, E. S. Brown, and on shipboard he met another millwright, Cyrus Walker, who would become an important bullwhacker in the future of the Washington lumber industry.

Captain Talbot and Cyrus Walker sailed north with a new ship, the *Julius Pringle,* in 1853 to survey things at Puget Sound. Both men suspected that Lafayette Balch and others were lying about the true conditions in the Oregon Country. Reaching the Sound, the pair were awestruck by the sight. Wherever they turned, their trained eyes saw timber . . . timber . . . timber . . . and on such a monumental scale that Talbot wondered if he were dreaming. The trees were unbelievable—Douglas fir, Sitka spruce, western hemlock, grand fir, red cedar—as thick as blades of spring grass and ripe for the cutting. The supply appeared endless, running to the distant peaks, although it was a moot question how you could ever get those far giants down to the sea and onto the ships. Those near the water's edge made it practical for mining with crude equipment.

It was a green El Dorado, the likes of which these Yankee lumbermen hadn't imagined existed anywhere in the world. They lost no time cruising the forests and looking for a proper place, for Henry Yesler was running ahead of them, and Captain William Sayward, another rugged Maine lumberman, was harvesting at a place called Port Ludlow. By longboat and canoe, they touched many points, at last settling on one called "Teekalet," meaning "Brightness of the Noonday Sun." It was renamed Port Gamble for a U.S. naval officer, although the implication seemed to be that they had some doubts about this remote wilderness.

The partners lost little time getting into production. Captain J. P. Keller, who agreed to build the mill, hauled up the machinery, boilers, engines, and store supplies from San Francisco while the men were constructing a bunkhouse, cookhouse, and store of Maine lumber brought aboard the *Julius Pringle*. Then, to avoid lost motion, they took on a cargo of piling and lumber from Yesler's mill before sailing south. By autumn the sawmill was slicing boards, only 2,000 feet daily at first, with the first sticks used to enclose the plant itself against the bone-chilling winds. But in its initial 12 months, the new plant cut 3,673,797 feet plus 64,000 shingles, 42,103 feet of piling, 223 masts and spars, and, significantly, foreign lumber shipments totaling 1,468,912 board feet for a year's gross business of $70,999. This allowed for capitalizing at $30,000. By the second year, the property was valued at $100,000, and within four years it was spewing 25,000 board feet per day and an annual output of 8,000,000 board feet.

By 1860 there were 25 sawmills rimming Puget Sound and any number of lumber ports, such as Gamble, Townsend, Blakely, Ludlow, and Madison. Many others were scattered throughout the Oregon-Washington region on the bays and estuaries and along the deep rivers. The schooners, brigantines, and barkentines swarmed to the crude docks, for there was no other way to get out the green gold except by sea. The bonanza was there, right enough, but as time went on, it became increasingly difficult to tame the big stuff as the timberlines receded from the water's edge. The lumberjacks cut the finest trees, the straightest and tallest, took only the best logs, and let the rest go, for the sawmills were always hungry, and the windjammers stood impatiently in the bays, their skippers cursing the delays and then cursing when the stuff was aboard the listing schooners.

"What in damnation ya tryin' ta do? Sink me?"

"Hell," retorted the mill man, "it's only lumber. You never saw lumber sink, didya?"

In the rush to mine Puget Sound, the lumbermen traveling up and down the coast overlooked one of the best timber stands of all, surrounding a lesser coastal bay named for Robert Gray, the American explorer who discovered the Columbia River. There were rumors, but little else. Curious about that part of the world, J. A. McGillicuddy journeyed by canoe out from Olympia in 1881 to have a firsthand look. Talbot had been amazed with Puget Sound; McGillicuddy, who had cruised a lot of timber sections, was stunned by the sight at Grays Harbor.

"I never thought there was so much timber in the world," he reported, echoing the sentiments of another early observer who described the place as "the noblest growth of fir, spruce, cedar, and hemlock ever found in the civilized world."

The trees were so thick and tall and close together, some of them growing as one, that McGillicuddy had to shoulder his way among them. Stands like "Old 21-9" on the Humptulips River were well beyond man's belief, the trees surpassing the giant coastal redwoods of northern California. A legend grew up about the twenty-eight sections of Old 21-9: Lumberjacks contended that so fine and rich and old were the trees that it certainly must be the mother township of all the Pacific Northwest Douglas fir region, that on the high winds its seed spread far and wide to create all the newer forests from deep into British Columbia to California. The legend is believable. Certainly, there were more fat trees surrounding Grays Harbor than even Paul Bunyan could imagine. McGillicuddy moved through the silent, beautiful woods, untrod by human beings for hundreds of years, like a man in a dream, unable to grasp his calculations. The timber ran 3 to 4,000,000 board feet to the 40 acres, over 50,000,000,000 board feet in all, five to 10 feet in diameter, and nearly 200 feet of sawlog lengths of forests nurtured for many centuries by cool, steady rainfall until the amazing growth ran 20,000,000 board feet to the quarter section (160 acres) and several hundred years old, just ripe for the cutting. McGillicuddy's excitement grew, his reactions similar to those of Dan DeQuille when calculating his first chilling estimates of the Comstock Lode. The great stands, all of similar quality and quantity, stretched seemingly forever through the rain-lashed southwestern Washington region and north to the timberline of the Olympic Mountains. It was a lumberjack's dream, as were the great silver strikes down in Nevada for the miner. This was the Big Bonanza, the Comstock of the timber world, and McGillicuddy dashed right out to report his findings, for he knew that soon the rush would be on.

McGillicuddy was dead right. William Anderson and Charles Stevens came that same year, setting up a shaky mill along the Chehalis River to cut Grays Harbor's first lumber. It was the tiny beginning of what grew into one of the continent's greatest industrial booms of all time. At about the same time, Captain A. M. Simpson was working his way along the West Coast with successful lumber operations in the redwoods and at Coos Bay. Simpson sent his aide, George H. Emerson, to explore Grays Harbor. Emerson was as speechless as McGillicuddy, immediately staking a claim to three hundred bonanza acres at Hoquiam before returning with the good news to San Francisco. The following spring Emerson brought up machinery and was cutting boards before autumn.

The Simpson mill was the first of any consequence in the area, but that kind of status wouldn't last very long. Lumberjacks couldn't keep their mouths shut anymore than could miners with a new strike. More sawmill men were coming as word got around, until in Grays Harbor's peak boom times of the twentieth century, there were half a hundred sawmills and wood-processing plants strung along the waterfront of the rough-and-tumble tri-cities of Aberdeen, Hoquiam, and Cosmopolis. Conditions for lumbering were never better than around The Harbor. Not only were there the rich green veins of heavy timber; the region was also slotted with many good streams, which swelled on the year-'round heavy rains—about 100 inches annually—so that the big logs could be sent a-whomping down to splash in the main rivers and deep bay, where the mills could handle them and ships of the world could gather with ease and safety.

Like Nevada's Comstock, the frenzy built up. Each new timber strike added to the mounting boom of The Harbor in what became a highball climax of three centuries of old-style American lumbering of the cut-out-and-get-out variety. In the six climactic years of the Golden Twenties, 1924–29, the annual lumber output by The Harbor's 12,000 or more lumberjacks and men was 1,352,-000,000 board feet of lumber; and in the twin peak years of 1925 and 1926, some 1,600 ships hauled off the bulk of a cut totaling 3,120,958,691 board feet mined from 2,000,000 mist-washed acres. In half a century, Grays Harbor shipped 30,000,000,000 board feet, that enormous volume equal to a year's production by all the sawmills in the United States. And in that small corner of the continent, everything came to a head in the sins of the past and the promise for the future. The Harbor became the lumber pace-setter for the Pacific Northwest region and indeed, for the nation. Timber ruled with a fist of tough, fibrous, pitchy wood, so that all else—the economic, social, and political patterns; the lives of nearly everyone—was molded, bent, and battered to its ends.

None of this timber production was easily won. Logging in the Pacific Northwest was a far hoot and holler from that of New England, the South, or the Lakes states, as the timber beasts learned upon drifting west before the turn of the century. The trees were whoppers, far bigger and heftier than anything they'd encountered, save for the redwood country of California. The big trees grew in rugged terrain—steep slopes and deep canyons that demanded all man and beast could muster in energy, musclepower, and ingenuity. The deeds were Herculean to meet the challenge; their god was Paul Bunyan, who did the impossible with ease and bravado. Paul and his Blue Ox weren't natives of the Northwest, but this was surely their kind of country.

The loggers were lesser extensions of the great Bunyan, a big, tall, tough, and ornery breed living in patterns and practices established centuries before in Europe and brought to the Atlantic seaboard with the colonists. For generations their unified cries were "cut out and get out" and "let daylight into the swamps." The beauty of the forests, the time it took to grow a great tree, which required only a few hours to fall, meant nothing to them, nor the destruction left in their wake. Even the British Crown tried curbing the lumberjacks in America from cutting every tree in sight, at least leaving the finest and tallest for ships' masts. But the pleas fell on deaf ears, and when the white pines had been cleared from New England, the lumberjacks took on New York and Pennsylvania, and even before the Civil War, they had invaded the rich stands of Michigan's famed Saginaw Bay.

On to Wisconsin! When that was done, they swarmed into Minnesota, where they felled and bucked trees by the millions, and what didn't get to the mills was destroyed in some of the greatest forest fires of all time, started by reckless loggers and settlers. They didn't care. There was always another forest over the next rise, the West stretched beyond forever, and the supply would last until eternity. As with oil and other natural resources, the lumbermen saw the supply as unlimited and everlasting. Not until they reached the Pacific shore, cutting trees with their backs to the sea and their tinpants soaked with salt water, did they suddenly realize that there wouldn't always be trees. It shook them to their calked boots, for whatever would they and the people do now? And for the first time, there was concern for the future.

As they moved West, the lumberjacks found scattered yet mounting opposition to their freewheeling days. Ranchers bucked them, since their clear-cutting eroded the soil and dried up the streams needed for the great cattle herds. Yet lumber was wanted by the very people who opposed them, and that was the ironic fact of it all. While people protested the destruction of the trees, the demand and the market were always there to spur the lumbermen forward. There were few controls over this wild industry. While mining and other forms of destruction of the natural environment weren't so readily visible, the forests were there for all to see and to love, and to write sonnets and songs about. The trees were living things of beauty and grace, tugging at the hearts of the people, even as they turned them into villages, towns, farms, and industries. Thus, the lumbering continued unbridled. America's first industry became its original industry of destruction—conqueror and tamer of nature on the frontier, if you will—but when time was running out, it became the first industry to begin turning things around, as its leaders realized that the end of the line had been reached, that "the old ways were dying in the rain," and that the wholesale destruction of the past couldn't continue.

Still, it was what is now called the hard sell. In the Pacific Northwest the lumberjacks had located the greatest stands of all, so the reckless practices went on. Now they were cutting at continent's end; when the last of this big stuff was down, they too would be finished—and so would the nation. Only a few visionaries were pondering long on that particular fact, but their words were most often drowned out in the roar of the donkey engines, the scream of the sawmills, and the dollars rolling along the skid roads. So they kept right on cutting, and the waste was unbelievable, with sawdust and wood chips piled so high that they threatened to bury the sawmills themselves. Runaway slash and forest fires blackened thousands of acres, and silt from topsoil slides brought on by heavy rains in the far hills, when whole logged off mountains fell away, turned the sea far from shore to a dismal brown, as hydraulic mining did in California.

This rain-washed timber demanded special challenges. Much of it would have escaped cutting had it not been for the inventive genius that developed new methods of logging and lumbering. East of the Cascades in the land of the pine, the lumberjacks continued using horses. There is still scattered horse logging to this day. But on the western slopes, horsepower was too light. The loggers turned to oxen, calling them bulls, and thereby developed a method that became the hallmark of western logging. Power—big power—was what they sought. Somehow that big stuff had to be gotten out of the brush. Unlike Grays Harbor, many streams were unaccommodating for log drives. There often wasn't enough snow to use sleds, as in New England; the land was deep, mucky mud; and the logs were far too heavy.

Some unknown timber beast figured out that by clearing a corridor through the tall trees and falling the smaller trees crosswise every few feet, you could skid the great logs along with the superpower of the bulls. Maybe it was old Henry Yesler himself, for he had the first skid road on the record books. Branches were cleared from the small firs that were half buried in the mud. Five to 10 yokes of 1,800-pound bulls were hitched to the logs in tandem, held by hooks, to the big sticks, dragging several at a time with beveled ends to avoid hangups. Amid the clanking of chains and the cursing bawl of the bullwhacker, this strange train or "turn" of logs snaked through the woods to the mill. Between the bulls and the big sticks ran the skid greaser daubing on a thick, gooey, foul-smelling substance made from dogfish oil to reduce the friction, which at times caused wisps of smoke to rise from beneath the moving logs.

The skid road worked, becoming an integral feature of logging operations—so much so that in the 1880s Malcolm McFarland even bored a tunnel through the hills above Westport, Oregon, to get his logs down to splash. Later, tremendous high wheels of 12 feet in diameter lifted the front end of the great logs from the ground, as a means of avoiding hangups, which brought yowls from those masters of profanity who ran the operation. Yet the bulls held an added advantage: the only kind of power brought into the woods that might end up in a logger's stomach. Accident victims could be eaten, and the loudest cry of all came from the lumberjack who bit into a yoke buckle in his beef roast. The loggers lived with danger. The traps and hazards, even at mealtime, were sufficient to drive him to the nearest saloon—close beside the right-of-way of his work. The logging district of the towns was therefore rightly named the skid road.

Bullpower left much to be desired, although it developed the logger's profanity to a high point of graphic, imaginative phraseology; the logs didn't move fast enough out of the woods to feed the hungry mills, now turning to steam power, and the skid roads could extend only a couple of miles back into the forests. This was the machine age. Down in the redwood country, one seafarer turned lumberjack did a lot of thinking about bulls and skid roads, and also about all that timber stretching forever yonder to the high slopes. It was a mighty long distance from the mills, and John Dolbeer's inventive mind had already brought many gadgets and innovations to the sawmill he and William Carson operated at Eureka.

Dolbeer was certain some kind of mechanical contraption could move those logs faster and more efficiently than the ornery bulls. Tinkering in his machine shop, he put together a crude donkey engine with a single cylinder, vertical boiler, and horizontal engine with drum. When he first tested it along a skid road, the cynical bullwhackers and hookers sat around on stumps, spitting tobacco juice and wisecracking. But Dolbeer had his engine fired up, there was a good head of steam, and in what seemed like swift seconds, he snaked a big log over the skid road in less time than it took to turn a bull team around. That day in 1881 would mark the end of bull team logging.

"Sam, I see you're finished," remarked a logger witnessing the sight.

The aging bullwhacker stroked his jaw, then snorted: "Well, the damned machine may be all right to git a log outa the pot holes, but when it comes to a skid road a mile long, ye kin jes' bet ol' Buck and Bright'll be in the game yit awhile."

By the following year, Dolbeer had his patent, which made him rich and endeared him to the hearts of the lumberjacks. Yet a certain color went out of the woods, for no longer would the bullwhacker's vindictive yells be heard shaking the trees and scaring the animals. Others had tested steam power, but not with Dolbeer's uncanny vision. The father of Archie Binns, a noted Northwest author, utilized steam at the Blakely mill once when he ran out of help and bulls. The mill needed logs, so Binns hitched a steam piledriver winch to a log and hauled it to the pond. Enough logs were snaked in this manner to keep the mill going, but the idea was then discarded, it being difficult to dislodge human beings from proven habits, so the credit goes to Dolbeer, who held the patent.

Within a few years, Dolbeer donkeys of various sizes and power were frightening wildlife in the Northwest woods. One thing followed another. The steam donkey developed ground lead logging, with the timber pulled by the turning capstan to the yard along the skid road and then the line hauled back to the far end by a team of knowing horses that could virtually find their own way. The line was set again by the choker on a log made ready by the sniper and the process begun on signal from the whistle punk to the donkey engineer.

It was the dawn of a new age, and no one could envision the changes. The donkey and ground lead system increased the volume of the cut and the output of the mills. The trees came down that much faster. In turn, logging operators and sawmill men increased their payrolls many times over. At the same time, the loggers were eying the railroads with all that great locomotive power, although the kind of railroading done on the main lines was far too fancy for the rugged forests. The woods were now on the verge of a machine revolution, and the next few years would see the greatest changes since Jamestown. Not until the 1970s, when one man could sit on his fanny in a warm, enclosed cab and manipulate levers and buttons, doing the work of an entire crew of loggers, would lumberjacks and lumbermen see such a vast change in life-style.

Chapter III

High Iron for Highball

An age of big things calls for big men. To invade a forest almost impenetrable, locked in the mountains deemed inaccessible, to successfully cope with the physical conditions which confront the logger of the Pacific Northwest, to develop the business ability to finance these business undertakings, and to market the product, calls for big men. The big man when fully developed is going to be one of the best types of big men, for he will of necessity be robust, vigorous, daring, bold, cautious, prudent, broad-minded and generous, for does he not work in Nature's own temple, the heart of the forest, drawing health, vigor and inspiration from his environment?

GEORGE S. LONG
1910
Pacific Logging Congress

"Always fall the tree toward the mill."

That was standard practice in the Big Woods. The land was steeper than the hills of San Francisco, which once also bristled with timber. Every inch gained was a saving in sweat, musclepower, and time in handling sticks so thick and heavy as to leave nothing to the imagination. The land not only stood on end; it was so mucky that the timbers wouldn't budge without a lot of muscular and verbal encouragement.

Thus from the beginning, grunt-and-groan logging was all on a decline—down the hillside, down the canyon, down the streams and rivers, down the flumes, down the trams and railroads, down the chutes, down to splash in the mill ponds at the far bottom—taking advantage of whatever gravity existed to skid those timbers and log sections. Power other than man and animal muscle was only what ingenuity and the inventive mind plus the present situation could create. The search for more and better power became a frantic movement as the high places grew steeper and steeper, and the good timber more remote from the sawmill.

You talked and coaxed and pleaded and cursed the logs from the woods. If there were a healthy stream nearby, you could float the sticks to the mill, provided the creek was deep enough so the big timbers could make it all the way and not snag up en route, creating a logjam that would clog everything like a freeway breakdown, clear back to Camp Five. Loggers waited for a good downpour in this rainy country or the annual spring freshets, but even as streams reached flood stage, heavy sticks hung up on the banks, never to be budged and eventually to return to the soil. Other giants were abandoned where they fell because they couldn't be moved from the canyons, like a fallen deer shot by a hunter who then discovered he couldn't pack out the meat. It didn't matter much, for time was money, and there was a never-ending supply of raw ore, green gold stretching forever and a day and lasting 'til Kingdom Come.

A logger had to be creative in making war on the rustic giants. Chutes were designed to bring the sticks down from the ridgetops. Later, some operators turned to flumes, which became more elaborate as time went on, some of them stretching for many miles of wheeling, turning, dipping, and crossing deep canyons on stilted trestles that were remarkable feats of engineering. Hundreds of miles of these amazing elevated wooden canals ran from the deep woods to the sawmills throughout the Douglas fir region, carrying billions of board feet of rough-cut lumber. Some companies established small mills in the woods for roughing out the boards, then sent them zooming down the flumes for finishing and shipping far below.

The sticks zipped along the fast-water flumes as on a toboggan run, miles an hour and gaining speed every 1,000 feet. The curves had to be judiciously banked so the logs wouldn't jump the track. Most loggers kept a safe distance when the sticks were coming down. Flumes ranged from a few hundred yards to 22 miles long near the Moyie River in British Columbia. Grades varied, some as much as 15 per cent. In Skamania County, Washington, there was a nine-mile flume 100 feet tall costing $10,000 per mile to build. Near Eugene, Oregon, the Lewis flume was legendary because of its many high crossings of a major state highway and the Willamette River during an eight-mile run. This was the scenic route, passing rival mills and through beautiful country to reach the sawmill at Pengra. The flume of the Bridal Veil Lumber Company in the deep Columbia Gorge was another noteworthy toboggan, dropping 1,200 feet in two miles to the mill near the river. The V-shaped slide carried lumber down to the planing mill and was able to handle 60-foot-long timbers. When there was a jam, it might take the loggers half a day to unsnarl the knot, but generally the flume was a cheap operation, costing $.10 per 1,000 feet in the early days.

Daredevil lumberjacks rode these logging freeways as they did everything else to get to and from the woods. The chutes were a dangerous way to go, but the lumberjacks lived constantly with violence, and the eagerness to spend Saturday night amid the bright lights offset the risk. Old-timers recall it as a "nice easy ride," although some could be hair-raising. During the week the loggers would save out the best cants for their tobogganlike trip to town, bobsledding out of the brush right down to the saloons and honky-tonks below in nearly door-to-door service. But flumes like the Bridal Veil weren't much for this style of transportation, although a small dog once rode a plank down the Bridal Veil clear to safety. But when a gallant timber beast tried it, he never made it to town for

Saturday night. On the first bad pitch, he sailed through the air and into the hospital.

Today, the flumes are all but gone from the Northwest lumber country. The only one of any size is operated by the Broughton Lumber Company on the Washington shore of the Columbia River, opposite Bridal Veil, seen a few years ago on the national television show "Lassie," with the famous collie riding the boards.

The railroad unlocked the main gateway to the green bonanza, and particularly a locomotive with the qualities of a foolhardy mountain goat. The pressure was on, for the lumbermen were running out of quality stuff close to the mills and the deep water. Rich veins of green gold were untouched on the high slopes and dark canyons beyond the rugged ridges, and as with the miners of the Comstock, the $64,000,000 question was: "How do you get out the timber?"

The Dolbeer donkey was a welcome sight, for its scream and clatter gave promise of a greater power source in the brush. However, for a long while the timbermen had been dreaming of the railroads for hauling the logs. Some built tramways, using wooden rails and horse- or bullpower. But regular railroads needed calm country, gradual grades, and sweeping curves, and the timber beasts had no time nor inclination to lay fancy trackage. The rod locomotives couldn't handle steep grades, especially with heavy loads. They had to stick to the flats and work near the sawmills. One timber beast felt he found the solution: They could build the lines in the treetops and thereby keep the grades nearly level.

Ephraim E. Shay, a rugged Michigan logger, followed in the shoes of John Dolbeer, believing that necessity was the mother of invention. Shay was compelled in the 1870s to cut his logging costs or quit business. He first built a tramway of maple rails, using horsepower, but then decided to try a "light locomotive" of steam power. He noted, as others had, that the rod drivers played havoc with the rails. As an alternative, he designed a geared locomotive, developed in his sawmill workshop with the help of his blacksmith. His strange, off-center steamer revolutionized the timber industry, for "Shay's Folly" and her imitators could go where the rod lokies feared to tread. In 1880 the Lima Machine Works produced three Shays, and eight years later there were some two hundred operating in the woods. And in the West, the Shay and her offshoots—the Climax, the Heisler, the Baldwin, the Willamette Shay—became a way of life, the kingpins of the on-end country. Given trackage, no matter how rough, the geared engine could follow the lumberjacks just about everywhere, like a mountain bighorn sheep, along a roadbed that was laid out by the devil himself, up steep grades and across creaking high trestles that no self-respecting engineer would dare take his pet locomotive and through lowlands and uplands and around curves so sharp that trainmen swore the headlight shone into the firebox.

It was natural that the timbermen take to the rails once the way was pioneered. They were historic allies of the railroaders, with a similar liking for big and powerful things. The lumbermen helped move the iron horse west, supplying billions of board feet of timber for bridge and trestle piling, ties and tunnel shoring, stripping the western forests to meet the demands of a long haul across the plains for the construction of the great transcontinental lines, while the railroad companies became heirs to the massive timber tracts given as rewards by the generous federal government. It was purchase of such rich timberlands from the railroad kings that brought the lumbermen back into the picture; and the epitome was reached when railroad mogul James J. Hill sold 900,000 acres at $6.00 an acre to the Weyerhaeuser Timber Company of St. Paul, Minnesota, bringing the big timber outfit to the Pacific Northwest, where it would become both a major force and a pace-setter in forest practices of the future, and causing Frederick Weyerhaeuser to comment with remarkable clairvoyance that this is "not for us, nor for our children, but for our grandchildren."

On the West Coast, the sawmills hummed to build the Southern Pacific system in California, across the mighty Sierra Nevada Range and north-to-south, while in the Pacific Northwest the sawmills of Ben Holladay, the West's transportation king—and later Henry Villard—clattered and screamed night and day, cutting ties and timbers to support rails bound for California. These railroads spread east and north, too, running on millions of wooden ties. In time the flimsy main lines became elegant and well-mannered, with a life-style that bound the nation together as never before or since—trains operated with pride, comfort, and safety in what I remain convinced is still the best and most sensible form of mass transportation ever devised.

The logging railroads were a different breed, the illegitimate offsprings of the proud mainlines and often disowned by the railroad purists because they were operated in such slipshod fashion, more on the ground than on the iron rails. They were as rough-and-tumble as the industry they served. There was no uniformed train crew with engineer and fireman in freshly laundered striped overalls, no linen on the tables of a fancy dining car, no plush private car at the rear for the president of the line. These were working railroads, and if the president of the line—i.e., the lumber or logging company—had a private means of getting about, it was likely to be in a flimsy speeder or handcar, much to the consternation of the locomotive engineer. Often the hoghead didn't know where the boss might be and could very likely meet him head-on around the next sharp curve. No small number of boss loggers were eliminated that way, for strict schedules were beyond consideration. If a pole train ran on time, it was the miracle of the age. There were many cases like that of the lumber king who was struck in his car at a train crossing, because for the first time within anyone's memory, the train loped by on schedule.

The timbermen slapped rails back into the wilderness without knowledge that it couldn't be done, dynamiting their way through brush and timber. All rules for accepted railroading were broken, and new tracks blazed piecemeal with scanty regulations. The railroads became the lifeblood of the industry, the leader in one of several revolutions to eliminate animal musclepower from the woods. In time, the logging highball became synonymous with the railroad highball, from which the term originated. Both were equally the most dangerous work in the world. Logging train crews operated poor, ominous equipment over soft roadbeds of worn-out track, carelessly spiked into weak ties or rustic sections of logs and supported by next to no ballast, for in shifting trackage from place to place, ballast appeared as something that could be done without.

Great 80-foot logs weighing several tons were snaked around thirty-degree curves on disconnected trucks and hand-braked down 10 per cent grades by nimble-footed train crews who danced from car to car along the tops of the logs and were ready to "join the birds" on split-second notice if the pole train took a spill. Many trains did jump tracks and plunge into deep canyons or ran away hell-bent down the grades, giving a final lunge over the bank to end with a sizzle in fast-running rivers. There were washouts, slides, and floods, and roadbeds that simply slid away without warning during the driving rains. Livestock on the tracks was another hazard. Many lumberjacks were killed working the trains. Lost fingers from the link-and-pin couplings became an occupational trademark similar to that of the shingle weavers. Just staying alive was the name of the game, with the railroad accident frequency outdistancing the high climbers, whose work was rated as among the five most dangerous occupations in the nation. In one year 22 trainmen were killed and 243 injured in Washington State alone.

In such an atmosphere, morale was bolstered by the rough humor of a hot foot, burning gloves, a hornet's nest in the engineer's cab, or dynamite tossed among the crew paused to eat their lunch. Yet the determined pot boilers and coffee grinders scaled the high ridges and toothpick trestles 100 to over 200 feet high, and crawled over dizzy switchbacks that would drive a man to drink. Usually the engineer kept a bottle of bravado in the cab, for while the general rule was no drinking on the job, this was a place where a little nerve calmer was sorely needed and a breach of the rules was sympathetically overlooked.

The lumbermen bought their equipment at bargain prices from mainline roads and streetcar systems. The conglomeration of obsolete running stock would today make a railroad museum curator drool with envy. Only the locomotives were purchased new and were the pride of the company owner, especially if he had more than one. What was pulled behind appeared as a worn-out circus train when a logging camp was on the move, including cast-off interurban and observation cars that were turned into "crummies" to haul the loggers back and forth. The strange accumulation ran on a variety of track widths, from narrow-gauge to not-so-narrow, standard, wide, and wide as hell—twenty-one inches to nine feet between the rails. Some outfits had more than one width of running stock, so made do by adding a third rail to take care of the wider or narrower trucks. It added merely one more facet to an operation which was so unbelievable that one old-timer remarked, "You can't lie fast enough to keep up with the honest facts."

A vast network of these stump-dodging railroads romped through the timber regions from northern California deep into British Columbia. Most anywhere you turned in the logging industry's great years, you found the rattling pole trains hauling astounding strings of cars heavily loaded with log tonnage, creeping at snail's pace along the twisting trackage. I recall in the early 1930s, when we first traveled the then-new Oregon Coast highway, stopping for a picnic lunch at a roadside pulloff near Coos Bay, where some of the first logging trackage was put down. Suddenly far below in the silent timber, we heard the deep rumble and clatter of a slow-moving pole train, a chain of around a hundred cars snaking for miles through the trees, being pushed, pulled, and shoved by steam locomotives fore, aft, and in the center, yet taking forever to disappear from sight. Before we headed on, two others rattled along the canyon right-of-way, headed for some distant mill. This was a remote but busy railroad, and a scene repeated many times every day throughout the Douglas fir belt.

Stewart Holbrook, the lumber historian, kept a record totaling 292 privately owned railroads he had ridden in four western states and Canada. He guessed that there were at least 200 others which he never rode. About 3,000 steam locomotives were in service in the woods during railroad logging in the Northwest, and at least 2,000 were active during the peak years just prior to the Great Depression. Over 300 timber companies owned their own railroads, averaging three lokies apiece, with two of them geared. The rattlers groaned over some 7,000 miles of iron and steel, the trackage expanding at a rate of 5 per cent a year. Average length of a line was 25 miles. Some hauled long distances, others only a few miles, but the general range was from 18 miles to over 100 miles.

Most of the Big Cut mined between 1900 and World War I came out of the mountains on rails, hauled by the geared Shays, Heislers, and Climaxes, and then through the flatlands by direct connected conventional Baldwins, Americans, Porters, and Vulcans. By World War I, 3,853 miles of logging railroads snaked through the north Pacific Coast timber country, over which chugged and churned 593 geared locomotives and 267 direct-connection types, hauling 8,000 sets of disconnected trucks and 5,500 flatcars. Ten years later, when 12,000,000,-000 board feet of timber went to the mills from Oregon and Washington, lumber's high iron had doubled its trackage to 6,164 miles of operable track plus 740 miles under construction. Rolling stock was up to 825 geared lokies, 316 rod engines, and 10,500 sets of disconnected trucks and 5,500 flatcars, about the same as a decade earlier.

The giant logging companies went for railroads in a big way. There is a picture, taken about 1940, of the yard of Polson Brothers of Grays Harbor, showing 14 fine steam locomotives in this single view, each probably representing a $10,000 investment. Timbermen were proud of their pot boilers, whether they owned a single snorting ornery hog that had seen far better days or a sleek new one, fresh from the Willamette yards. Indeed, when one fast-talking lumberjack got himself the job of operating a new locomotive for Schafer Brothers and proceeded to wreck the $10,000 steamer the first trip down to the mill, he took off like a frightened rabbit. Months later, Albert Schafer ran into him and asked where he'd been.

"I been running," the fellow replied.

The loggers ran their own shows with the rails, even building fantastic inclines and declines to bring the logs off the high ridges on grades as steep as any of San Francisco's cable car hills. Heavily loaded cars were precariously winched down grades of 45 to 70 per cent, held by steel cables to the steam donkey and counterbalance weights. If a line snapped, all hell broke loose. Once it happened on a spur of the old Punk, Rotten, and Nasty in the Tillamook Burn country. The loaded log car careened down the sizzling grade, took out the mainline below, and sailed on across the canyon to a magnificent crash landing on the opposite side.

Many loggers refused to ride the inclines, considering them "too dangerous." They had their doubts, too, about some of the flimsy-appearing trestles, preferring to pick up the train on the other side than take in the view from 200 feet in the air. One train crew worked out a system for avoiding a particularly ominous-looking trestle by setting the lokie in motion, leaping off, dashing through the canyon, and catching the hog on the other ridge. Another crew utilized two lokies, fore and aft. They'd work from the rear engine to push the cars across the span, then stop the train, hike through the canyon, and resume the trip from the forward engine.

Certainly the trestles, thrust above the treetops, gave reason for doubt, yet they were amazing victories of engineering. Most of them were understandably of wood, sometimes containing enough timber to build a small town. A quarter-million board feet might go into a single trestle. Yet trestles were cheaper than fill and less exasperating. When one railroad bridged a bog area, the filling seemed endless. Each time they ran a train across, the roadbed acquired a swayback. In went more fill, the track was jacked up to grade level, then another train came along, and the whole operation had to be done again.

The trestles were knowingly not built to last, expected to fall down "after the last train passes over it," and therefore loggers and train crews had their rightful reservations about their safety: made of toothpicks and bean poles, they claimed. Among the highest was across Cedar Creek near Sellack, Washington, towering 204 feet above the water and consisting of 508 piles, which would stretch seven miles if laid end to end. Yet the grandest, engineered proudly by Walter Ryan of Weyerhaeuser Timber Company, bridged Baird Creek on the Longview logging line. The spectacular span contained 400,000 board feet of treated timber plus 200,000 feet of untreated piling and members, fitted together for the 1,130-foot span like a giant Erector set involving thousands of pieces surrounding a timber arch, which was indeed a work of art in any man's language.

Yet such bridge work was expensive to construct. When you had to do it several times, it became almost too much. McCormick Lumber's first trestle was knocked down by a spring flood. They decided to move the operation, so rebuilt the trestle to move the camp and equipment, then tore it down. Suddenly the super discovered in horror that two locomotives were left back in the woods. So cursing crews built the trestle a third time to bring out the stray lokies.

Loggers' kids found the towering spans glorious playground equipment for staging races, contests, and for initiating greenhorns into a "club" or "manhood." Parents warned their offspring about playing around old trestles, which were as dangerous from rotting and near collapse as those where the trains still ran. But life in the woods was surrounded by risks, death to the loggers was often instantaneous, and many a boy of this tough breed found thrill in rough-and-tumble activity. If trestle racing wasn't in the cards for this day, there was always tree riding—a hair-raising downward descent in the top of a tall fir felled by one of your friends; then you'd trade off and do the same for him.

The lumberjacks depended upon the logging railroads and their running stock to solve any problem, meet any challenge, whether it was wrestling the big sticks off a remote ridge, pounding piling for a new trestle, or serving as a skidder. They used and misused the hodgepodge equipment, and when something broke, they just patched it up again. Roadbeds held up amazingly well, considering the terrific tonnage, which threatened to spread the poorly spiked iron—and many times did. Derailments were commonplace; engineers claimed they often ran on the ground. When one logging line had its rails straightened, the ride was so smooth that a lumberjack shouted: "Jump, everyone. We're ridin' on the ground." As one lumberman put it: "It isn't the curves that bother us; it's straightening out the track we've got." Only a sense of humor and a dash of fatalism kept many of the lines running at all.

The lokies took on all types of duty, from makeshift donkey to fumigating the bunkhouses with live steam. Sometimes lumberjacks didn't bother fitting the sticks onto the trucks or cars, but made a skid road out of the railbed, dragging the logs behind the engines. This action did nothing to improve the roadbed. On icy mornings on the steep slopes, loggers sent the sticks down to the mill under their own weight, by gravity alone. All it required was a slight shove. At one camp an unscheduled log got away on the ice, heading downhill with a lokie coming up, and gaining speed with every second. An alert lumberjack, seeing disaster if it met the engine head-on, leaped aboard with a powder box of dirt and sand, trying to slow the runaway in time. After a mile's wild ride, the log slowed sufficiently for him to jump off to spread sand on the ice ahead of the timber, which at last came to a halt. But it was a ride he never forgot.

Logging railroads moved with the lumberjacks to each new show. The roadbeds were laid out by trial and error until operators began hiring trained construction engineers. The loggers had a name for them, as they did for everything: scenery inspectors. Operators hoped these specialists could cut down on the delays, losses in men and equipment, and the dangers. But the loggers, especially the woods boss, who believed he could lay out any damned railroad, scoffed at the college-book larnin' of the experts. The scenery inspectors were charged with locating the cheapest, fastest, easiest, and safest routes to the high timber, with safety bringing up the rear. Sometimes they were forced to build from the operation out to the mill, across on-end land that offered unprecedented challenges to their engineering skills and some remarkable results in cliff-hanging rails that ran through the clouds and into canyons deep as Hades itself.

There was a passionate longing for balloons to loft everything over all this, or hang the rails from skyhooks, and if it hadn't been for a useful tool called dynamite, the way would have been much tougher. But the loggers weren't to be denied; when one landowner refused to allow an outfit to railroad across his property, the timberman bypassed the stubborn rancher with a spectacular railroad cable system down a 70 per cent grade on his own holdings.

The lokies were fired with cord wood from the forests and scraps from the mills in what might be called the first recycling or full use of manufacture waste, although the loggers didn't view it that way. The stuff was merely handy, practical, and made good sense. When the operation moved, the railroaders simply hauled off the camp—bunkhouses, cookshack, and all—right behind the scenery inspectors and their gangs laying down the track. The bunkhouses and cookshack were often left on the cars, shuttled from siding to siding in what were undoubtedly the first mobile homes. This wasn't necessarily the best location, however, for the cook and his artistry, especially if he were baking when a train went by. One engineer got pie on his face rather than in the sky the day he put his lokie through the wall of the cookhouse.

When the camp moved on, logging was done in that area. If not on wheels, remnants of the abandoned logging camps were like those of the worn-out mines, left to the squirrels and woodrats, with the brush taking over and the new forest growing up around the shacks, eventually crowding out the sagging buildings until they collapsed from the pressure of the trees and the rotting away from the rain. Little remained after a few years, save for a single stout cabin or two, a few whiskey bottles, and a rusting snoose can. Sawmills and their towns, too, were likewise abandoned to rusting splendor when the green gold ran out. Unlike the gold and silver towns, an old logging camp is hard to find today, for deterioration was much more rapid due to climatic conditions of the forests over that of the dry desert.

The lumberjacks traveled the rails to the next logging show or to town for the Saturday night howl as they did the flumes, straddling the sticks or riding in the jerky caboose in their cutoff high-water blue jeans, wool shirts and socks, red flannel longies, plaid mackinaws, calked boots, and red flannel hats with crumpled brims perched jauntily on their heads. Mostly they were young, but already battered and bruised by life. The enclosed caboose held the welcome comfort from a potbellied stove putting out heat to thaw their bones, on which brewed a pot of the stoutest coffee in all the Big Woods—black as the hole of Calcutta, which if spilled on your jeans would eat a hole right through your leg. If there wasn't a "Millionaire's Special," they'd grab the nearest handcar or hike down the track, but the best way was to talk an engineer into making an unscheduled run. However, when a half-dozen would-be celebrants sold a steep-grade hogger into taking the lokie to town without a counterweight, it cost them their lives.

All year long, day and night, the bindle stiffs mingled with danger and death, as do front-line troops in any war. Life was giving them few of the world's goods, and their dreams were pie in the sky. Mostly they were transients on the move from job to job, having no families or permanent homes, forever going with bedroll and ax to that latest logging show over the next rise. It was all they knew, and despite the wet and cold, the danger, the high risk of a bloody death from splintered timbers and whipping cables, runaway trains, and crashing logs, there was the challenge of destiny wrestling those big ones that God had put there long ago, surface mining in good clean air and the overwhelming surroundings of flora and fauna in this wonderful Northwest country. That they were destroying the very thing which attracted them didn't occur to the loggers. It was a job, something that needed doing, and it brought a paycheck at the end of the week; or in the case of Pope & Talbot, a fist of silver half dollars, for it was company practice to pay with the four-bit pieces daily or weekly. At Port Gamble, the heavy treasure arrived by steamer from the San Francisco mint, and the local postmaster, D. L. Jackson, hauled it from steamer dockside to his safe by wheelbarrow, right up the street unguarded and with never a holdup. It was recognized at once in the saloons or shops, if a fellow doled out silver four-bitters, that he worked for Pope & Talbot; and one employee sent to Alaska for three months on company business collected 900 pieces of silver when he asked for his back pay.

Seldom a week went by without a buddy killed or maimed by a rolling log, a falling tree, a giant splinter run through him, or a whipping cable slicing him in two. From the time he got out of bed in the morning, the lumberjack never knew if he'd make it through the day, for even in the dining hall you might get a fork through your hand grabbing at the last chunk of steak. Lumbering remains a dangerous occupation today, despite the push-button equipment and the safety regulations. I recall attending the 1971 Pacific Logging Congress where one particular speaker about uphill logging reminisced that if anyone had ever been in the path of a great wild log coming down a steep hillside and wondering whether his churning legs would get him clear in time, he knew what it was to live with danger. Everyone in the room hooted with laughter, for in mod-

ern vernacular, the fellow was certainly "communicating." Many of the dele-
gates from all the West bore the scars of their trade and were minus fingers,
for although their pockets now jingled with coin of the realm, they'd come up
the hard way.

Highball logging wrought the greatest havoc to man and timberland, and the
railroads with their long shrill twin blasts—the signal of highball—made it all
possible. The sawmill half of the industry was equally dangerous. The flimsy old-
time sawmills were death traps from flying timbers, faulty saws, high speed,
carelessness, and a lack of safety regulations. Mill owners gave little thought to
safety, nor did many of their employees, who likely came out of the woods and
accepted danger as going with the job. One old-time labor leader, a man of few
but effective words, described safety of the early mills as "nil." Fingers and eyes
were the most vulnerable, especially among the shingle weavers, who could
quickly identify each other from the marks of their trade. Bad lighting played
a significant part in tangling with a whirring saw. When one worker ran his
hand through a saw, he lost all the fingers and passed out.

"How in hell did he do that?" asked the bossman, rushing up.

"Why, just this way," replied a young buck, and demonstrating, did the same
to his own hand.

Accidents were often bloody and spectacular, especially as the highball picked
up steam in both woods and mills on the advent of the high lead and the great
circular and band saws. A circular saw running at high speed without adequate
guards could throw wood and splinters with the force of a guided missile right
through the side of the building, or take off a man's head. A small chunk could
claim a man's eye or his scalp, while a huge javelinlike splinter could pierce a
fellow's midsection, killing him outright. Plenty of blood was spilled on the floors
of the sawmills. Boilers exploded and slipping logs fell on workers, crushing arms,
legs, and chests. Sometimes the fast-traveling log carriage careened off the end
of its track and through the mill wall, carrying the operator to his death. Circular
saws broke loose while spinning at high speed, sailing through the wall or the
roof into the wild blue yonder, and everyone hitting for cover. If the running
saw struck a railroad spike or iron bar imbedded in the log, all hell broke loose
from metal shrapnel flying in every direction. During the wild strikes and
Wobbly activities of World War I and the 1920s, this was a favored method
of sabotage, as was the dynamiting of mills and railroads.

The great band saws, ribbons of steel 60 feet in circumference, 16 inches
wide, and running at over 100 miles an hour, were another matter. When one
broke loose at the welding line, or from metal fatigue, the mill became a hall of
terror from a monster of whipping steel, like a giant clockspring unwinding in
all directions. Crews dived over the side, ran for their lives, or huddled beneath
the pilings. Harry Jenkins, manager of the pioneer Jones Lumber Company mill
in Portland, found himself once doing a death dance with a whipping band saw
on the make. The quivering blade came at him, faded, and came again, close to
shaving his whiskers. Jenkins could only stand his ground until the monster
danced off in another direction, a thing alive and uncontrollable.

Rampaging saw bands were known to go through the sides of the building and
into the yard. At other times the saw would roll into a corner and crouch there,
like a coiled serpent, ready to strike. The band's energy was stored in the coiled
blade, and the men feared to touch it. One time, sawmill crews bravely blocked

up a blade, cementing it permanently into the corner of the mill. It was touch-and-go, for the industry had no daring demolitions experts. Energy was still stored within that blade quivering in the corner. It could take off without warning, and nobody wanted the job of getting it out of there. There was far less danger in prowling the dives of skid road.

Accidents were often the kind that ripped at a man's guts, either in the woods or at the mills. A man crushed by a rolling log or a falling tree wasn't a pretty sight. Neither was one who'd fallen beneath the wheels of a logging train. There was little left for the cemetery. Harry Jenkins saw a worker fall into the gang saw, which shred him to ribbons. Years later, it still made him ill thinking about it. The coming of the high lead, coupled with the highball, added substantially to this most dangerous of all occupations, beginning with the high climber or high rigger, who topped the great trees for the spar poles and set the block at the top. No circus act ever equaled the high climber. He served like a steeplejack or stunt man with no safety netting, sawing the tops of the stately fir 150 feet above the brush, then riding out the storm of the whipping trees as the tops broke loose to crash to the ground. The danger was acute at that final second when the top broke away, for if the tree split, the logger could be cut in two by his safety belt wrapped around the tree.

With the main line rigging running through the 1,800-pound block at the top, plus a haulback line from the donkey, the great timbers could be lofted from the woods, the closest thing yet to the loggers' dream of a skyhook. The hefty logs sailed freely through the air, avoiding ground hangups on stumps and underbrush and making sniping unnecessary. The high lead almost doubled the output, to the glee of the bossman and the sawmill operators, for the mills were forever hungry. The bull of the woods offered bonuses based on volume, and this encouraged recklessness; everyone from choker setter to the trainmen had to be nimble-footed to get out of the way, for signals were given before the hooker was clear. Every second counted: The more volume, the more cash money in the bossman's blue jeans. The slow man cost all the crew, so he either quit or was maimed or died beneath a four-ton log. The old-time lumber kings sacrificed countless lives getting the logs at high speed down to splash. Some, like an operator in Yamhill County, Oregon, had widespread reputations as killers who didn't give a damn. Moreover, many small operators ran with poor equipment, including weak and damaged rigging, which added to the dangers. Few bothered to check rusting cables regularly, so they snapped without warning, hurtling through the sky like Paul Bunyan's bull whip. All told, the greedy operators didn't care; the lone vagabond logger was dispensable, so long as you got out the logs and made it pay.

The lumberjacks and their bosses were equally careless with fire, for the same reasons. Stumpage was cheap; if you destroyed one stand, another was over the next knoll. It was worth the gamble in hot, low-humidity weather to keep operating rather than shut down or go to hoot owl (sunrise) logging, when the humidity was higher. In low humidity, friction from two logs rubbing together could explode the forest in seconds into a raging holocaust. Hundreds of thousands of black acres were touched off by damned-if-I-care operations, including the greatest forest holocausts of all time. It took many decades and much teeth gnashing to bring about laws that would automatically close down

the Northwest forests in extreme fire danger. Those laws changed things, so
that there would be suits over fire responsibility, brought by neighboring timber
holders against careless offenders who started conflagrations that spread to
other lands. The new laws, forest codes, and resulting suits helped curtail the
recklessness, but there was always that lone maverick who had to move just
one more stick before shutting down, or didn't close down at all, and away
went another fine virgin forest which had taken hundreds of years to grow.

Sawmills too were plagued by fires, especially in low-humidity times, often
blamed on the traditional wigwam burner, which became symbolic of an age
and a way of doing things. Early mills, because of the massive waste coming
largely from the wide kerfs of the circular saws, were almost buried by their
own waste in sawdust and chips. "Sawdust factories with a by-product of
lumber," one wag described them. The stuff was dumped into rivers and
swampy lowlands as fill and spread on the muddy streets and walkways of
the sawmill towns. Someone came up with the idea of a huge burner shaped
like an Indian teepee, adjacent to the mill. The wigwams became a familiar
landmark in the Northwest sky, the smoke from the burners visible for miles
before you could see the sawmill, feeding day and night on the waste carried
by a conveyor system from the mill. There were hundreds of wigwams scattered
on the coast, in the mountains and valleys, along the waterfronts of the cities
and towns—the marks of industry. There was also the towering 200-foot all-
brick stack of Pope & Talbot's mill at Port Gamble, finally torn down in 1972
after long being a familiar navigation point on Puget Sound. The burners,
several stories high, sent a constant column of smoke into the air along with
sparks and flaming bits of wood. Despite its obvious drawbacks, these unique
furnaces were nevertheless the solution to a problem that had plagued the
lumbermen for many years. A smoke pall from the mills and the logging slash
fires hung continually over the lumber country, an accepted part of the Northwest scene long before there was such a word as "smog." This was especially
true in the fall of the year, when logging crews were burning slash; and few
complained of waste or air pollution, since this was the region's foremost
industry and an accepted way of life.

At the outset of World War II, the wigwam burners suddenly became a
menace that would help the enemy. When the Japanese attacked Pearl Harbor,
the frightened Pacific Slope was blacked out for fear that an invasion force
was lurking off the coast. Then military and air raid officials were horrified to
discover that the Northwest was lit up like the biggest Christmas tree of all
time by hundreds of wigwam burners glowing in the night, which could be
easily spotted along the open coastline, far at sea, and could guide enemy
planes to inland targets.

In every mill town, defense wardens went around dousing the wigwam fires
over protests of the lumbermen. When the blackout was temporarily lifted, it
took 25 barrels of crude oil to reignite the burners. It was off again, on
again; blackouts and alerts came at any moment during the war's critical
early months. The sawmills couldn't operate without the wigwams, and crude
oil was a costly item, if you could get it at all. It was a trap, for lumber is a
leading military necessity, and the government was placing orders as never
before. The sawmills must keep running, or the war might be lost.

An urgent appeal was made through the huge and powerful West Coast

Lumbermen's Association for an idea that would keep the wigwams in use, yet could cut the fires quickly in an alert. A heavy dousing of water made the burners difficult to rekindle and might injure the interior brick linings. Finally a sprinkler system was devised, placed 20 feet above the fire cone, to create a fine mist or fog in the upper portion of the burner. As it settled upon the fire, the flames were squelched in two minutes' time, the sprinklers were turned off, and when the blackout ended, even a long one, the fires could be quickly rekindled on the glowing coals.

Thus this helter-skelter industry flourished, in haphazard fashion from crisis to crisis, in war and peace, good times and bad. The cruisers staked out the green gold sections, the high climbers scaled the towering firs, spruce, and cedar, and the fallers sent the trees tumbling to the rain-soaked forest floor. The buckers sliced up the great logs, taking only the choicest, the chokers and hookers got the cables around them, and the whistle punk standing on a stump gave his signal. The heavy timbers went crashing through the woods like prehistoric monsters, yarded along skid roads or gouging out the brush and tiny oncoming generations of trees, to a river or loading dock, where they traveled by rail, by flume, by chute, or on the raging torrents of a wild stream by the millions, so many matchsticks floating down a gutter on a sudden rain, down to splash in the mill ponds or bays or deep rivers. Then they were boomed up to the final trip to the sawmill, to be halved, quartered, and sliced into many grades of lumber for the massive consumption of the nation and the world, for everyone needed lumber, and it was the handiest thing around.

Chapter IV

Kill the Fiddler with Him!

"What would you do if some fine morning you came down to your mill at 4:45 A.M. and found your entire crew not only at their posts but had—without orders from your office—started work at midnight and had manufactured 225,000 board feet of lumber?"

"I'd give up drinking."

FOUR L BULLETIN
1925

When the woods were shut down and during the annual Fourth of July vacation, the timber beasts headed for the "outside." "Town" might mean anything from the little clapboard village surrounding the sawmills they'd been feeding for the last six to nine months to the sprawling lumber ports, where there was far more to see and do—a larger assortment of deadfalls into which a fellow might upend himself.

Pockets bulging with six months' pay, the loggers rode the wild flumes, the rickety railroads, horse-drawn wagons, or hiked the muddy foot trails from the brush. When they hit the lumber towns by the thousands, the scene was much like that of the cattle centers of Dodge City and Wichita.

The loggers came to blow 'er in amid the bright lights and excitement, to "get their teeth fixed," as some of them put it, after long, dreary months isolated in the damp and lonely back country. They headed at once across the deadline to the skid road district, the toughest and bawdiest section, where there were no holds barred and a man made his own rules with his fists or whatever happened to be handy for swinging on another. The lumberjacks mingled, drank, argued, and fought with other working stiffs—sailors fresh from months at sea, cowboys off the eastern ranches, fishermen out of the seining grounds, miners from the mountains, riverboat men, waterfront stevedores, and laborers of the industries. The skid road was a western institution; it belonged to the men, and amid its dives, saloons, and bordellos, the bindle stiff found release from all those piled-up emotions and drives that had been bugging him for the past half year. Since most loggers were single and transient, this was part of their style, their way of life. It didn't matter much if a jack blew all his poke at gambling or lost it to some harlot, for when it was gone, he headed right back to the brush for another long siege of highballing timber, where he might be killed or permanently crippled. The perils of the skid road, therefore, held little fear for him, as destiny had dealt him a fateful hand; he was certain to die young, and tragically, deep in the Big Woods.

In that wilderness, the timber beasts hungered for escape entertainment when they weren't too beat from 10 to 12 hours' wrestling logs just to grub down and tumble into their bunks, to begin it all again early the next morning. In that cold and miserable forest, leaning hour after hour into a heavy crosscut while balancing precariously on a thin springboard, or at a dozen other grueling tasks that took the sap out of a man, the lumberjack felt life offered him very little. However, there were times—on Sundays, maybe—when other cravings pressed against him. A good fight, wrestling a half-tamed bear for a pot of coin, contests at wood chopping, tree falling, high climbing, burling, bucking, ax throwing, and dice and card games were ways to pass the time and unwind from fighting the brush, dodging lines and splinters, and scrambling hell-bent away from a wild log coming your way. These activities became the foundations of the timber carnivals of later years, the first one being the Grays Harbor Splash, held alternately at Aberdeen and Hoquiam. The YMCA also invaded the camps to help pass idle hours and keep the boys from drifting down to the saloons on the perimeter. No liquor was allowed. It was a tight, rugged routine for the wage slaves, and sooner or later came the day when the logger, restless or soured on the outfit, pulled the pin, cashed in his time, and swung aboard the next outbound logging train.

"If you don't like it here, mister, clear out. It's a free country."

It didn't take much to move. Like the cowpoke, the logger's gear consisted primarily of his calked boots, ax, blanket, change of heavy woolen socks, and the skills he'd learned in the woods from Maine to Michigan, maybe into southern pine, and now here on the Coast. He stashed his gear in the walk-up room of a skid road hotel or a place like "The Men's Resort" at Fourth and Burnside, Portland, a haven for footloose roamers, with baggage storage, mail privileges, rest benches, and a lecture platform. He paid the hotel rent days ahead, for insurance, patronized a nearby bathhouse, got a haircut from a lady barber, and turned a goodly part of his roll over to a trustworthy bartender or shopkeeper for security. In Aberdeen, a banking business grew from a waterfront cigar store and magazine stand where loggers put their pay. Word got around that it was safe there, the proprietor being tough but honest. On Saturday nights his till might hold $100,000. He wouldn't give up a bankroll while the owner was drunk, despite angry threats on his life. He looked out for the loggers, and they were grateful.

The boss loggers and sawmill operators often observed that they had three crews: one coming, one working, and one going. Job sharks kept the men moving from camp to town, and back again, charging $1.00 to $5.00 to locate employment, in an arrangement with the bull of the woods. Poor grub, long hours, hard work, miserable sleeping conditions, little time for fun—these kept the men dissatisfied and on the move, back to skid road to blow their pay for the benefit of the saloons, railroads, hotels, restaurants, and dives in the constant movement of dollars. If a man got too "stakey," it was best to "let him drag his time," for the only sure cure for bunkhouse fever was hard liquor, a brawl, women, cards, and gambling.

The system kept money in circulation, for sure. Only Simon Benson, one of the lumber kings, tried to beat the pattern of the Fourth of July whoop because he lost his crew to a big drunk and a lengthy hangover, right when summer logging was at its peak. Benson hired Scandinavian crews, who quickly concluded that the Fourth of July was for the sole purpose of getting drunk. But a couple of weeks before the Fourth, Benson shipped barrels of whiskey to his camps, with drinks on the house. The Swedes had their celebration, stayed in camp, and when the Fourth came around, they were content to keep working.

"Sporting House Kelly" on the lower Columbia brought in a covey of women to run his cookhouse—not whores, but gals of good moral standing. Actually, it was his wife's idea, against his better judgment, in hopes that having the women about would make the loggers more content. Maybe Mrs. Kelly just wanted company. Kelly warned the crew to keep the language pure in the dining room, but the loggers were forgetful and the gals complained. One evening, brandishing a pickhandle, Kelly leaned against the cookhouse door frame. Then he roared:

"I've heard you goddamned lousy timber beasts swear and use dirty talk at the table. The first one of you bastards that makes just one foul dirty crack, I'll kill the goddamned sonofabitch right here. . . ."

The horrified women fled the hall and the camp. Kelly looked after them, puzzled, then growled to his wife: "There—now you see, don't you, that women don't belong in camp?"

Along the skid road, the loggers played as hard as they worked, in a big carnival of side shows and main attractions—saloons, bawdy houses, tattoo parlors, shooting galleries, outfitting stores, hiring halls, greasy-spoon cafes, penny arcades, a mission where you could get free soup, doctors and dentists' establishments, including the notable Painless Parker, smoke shops, and maybe a secondhand bookstore or two. Believe it or not, loggers were great readers. A place displaying Oriental goods meant a Chinese lottery, and the sign of the black cat indicated the hangout for the Wobblies. The pattern was much the same whether it was Seattle's Pioneer Square, where the first skid road came into being, Vancouver's Cordova, Spokane's Main, Portland's Old North End, Everett's Hewitt, Aberdeen's Hume, or the Humboldt Bay area of Eureka.

The High Lead . . . The Saginaw Rooms . . . The Cookhouse were popular names taken from the loggers' own colorful vocabulary. All beckoned to relieve the lumberjack of his hard-earned pay, one way or another. Hustlers, pimps, sharpies, strumpets, and crimps prowled the streets and walkways paved with sawdust from the mills, steering the boys into the bushwhacking traps. Yet each skid road had its peculiar characteristics, and the loggers were amused by their own cleverness of detail and their unique way of life. Aberdeen's red-light section was known as College Row, with places called Yale, Harvard, and Columbia. Seattle had its box houses, one of them the first theater of John Cort, which later became Joe Backer's famed Our House. In Portland it was Gus Erickson's long bar and grand saloon, and also the astounding Paris House, while in Aberdeen it was the Humboldt.

While loggers liked trappings that were rugged and utilitarian, they seldom exhibited enthusiasm for the ornate plush, crystal, and brocade known to the Comstock Lode miners of Virginia City and San Francisco. The lumber kings built their own rustic palaces, such as Cyrus Walker's Admiralty at Port Ludlow, for there was a distinction now of logging and lumbering as separate operations. The real fortunes were being garnered making little ones out of big ones, not from gutting the forests. The loggers were far too practical and earthbound, two-fisted rowdies who drank their coffee black and their whiskey straight—powerful rotgut that seared a man's stomach and put hair on his chest, or took it off if applied externally. The dry years didn't curb them much. The Pacific Logging Congress decided in the arid midtwenties to escape from this terrible calamity by meeting in Vancouver, British Columbia, which was wide open. The first night those loggers tackled snake bite like it had gone out of style—which it had, south of the border. It was a motley collection of lumberjacks showing up next morning for the opening session. The bleary-eyed first speaker wobbled to the front, missed the rostrum, and fell flat on his face, snuggling down on the carpet for a long winter's nap.

"Leave him lay," shouted those who were sober enough to see the action. So the chairman called the next speaker, who picked his way over the prostrate body. He knew nothing about his subject, but read a long, dry paper, which nearly put all the audience to joining the birds. When it was over, an old Scot shouted: "That was a damn poor paper and it was damn poorly read."

These boys had to have elbow room, and space to swing an ax. The Humboldt's floors were eighteen inches thick, but needed replacing every six months, chewed to splinters by calked boots. The calks also worked on the bars till they were virtually falling apart. The day would arrive when signs "No Calks Allowed" appeared in the lumber towns, and when that time came, the glory days were over. Saloon furnishings also needed to be thick and heavy to withstand the rowdiness and skull-smashing brawls that went on there. Hot-tempered loggers attacked challengers with anything from a whiskey bottle to a cant dog, or the hefty leg of a chair, and completed the job of changing the fellow's looks by scraping his whiskers with the corks. A sound axhandle was a fine weapon. Finally, the powerful timber beast would lift his rival high above the floor, spin around, and send him crashing against the backbar or into the tables, smashing glassware and furniture if not the logger. Giant bouncers gave similar performances, able to handle three hundred pounds with ease, and such a ruckus, with the victim lofted overhead, gave rise to that classic request:

"Don't waste him; kill the fiddler with him!"

Brawling began almost immediately on opening night at the Palm Dance Hall in Aberdeen, a sprawling dive that had five bartenders, four burly bouncers, and 40 girls. One battler grabbed a scantling with a spike on the end, swung it around, and scalped another logger, with fatal results. Off to a bad start, the Palm nevertheless enjoyed a long run as a favorite play center at Grays Harbor.

Some of the logging saloons became as legendary as Dodge City's Long Branch, Virginia City's Delta, and the Bella Union of San Francisco's Barbary Coast. The loggers never patronized the Barbary Coast joints to any extent, save perhaps out of morbid curiosity. San Francisco was too cultured for them. They liked Eureka, near those whopping big trees. But the lumber nabobs who ran the mills and held the fat bank accounts were attracted often to The Bay to enjoy the opera, vaudeville shows, and museums. Their wives could spend small fortunes on the latest styles to take back to Portland, Tacoma, and Aberdeen, where they rode through town in fancy rigs, dined at plush hotels like the famed Portland, and served tea in their rose gardens. One of these lumber kings, as a lonely old man with his time running low, hung out regularly at the Thalia on the Barbary Coast, buying drinks but never touching them and chatting with the waiters and girls, who came to accept him as something of a fixture about the place. Occasionally he jotted down their names in a little notebook when they were particularly kind to him.

"Thanks so much," he said. "I'll remember you in my will. I'm a wealthy man, you know."

Nobody took him seriously. Many strange characters and eccentrics frequented the Barbary Coast joints; and any number were liars, under aliases. But when the lumber king died, all the girls and the waiters, and even the proprietor, were remembered substantially in his will.

Seattle's Our House ranked as a place of rugged distinction around the turn of the century. Our House was easy to locate, fronted by great barrel heads painted on the outside walls, each one lettered, "Only straight whiskey for Our House patrons." There was also a huge scene of a European tavern. Giant pillars flanked a grand mirror behind the mighty bar of green tile and panels containing intricate carvings. Upstairs gambling rooms provided for 25 games going as long as customers held out. Our House ranked as skid road's most illustrious hangout, yet despite the fact that Joe Backer founded the place for loggers and the tradition was carried on by the Landwehr brothers, a startling sign met customers at the door: "No Calks Allowed." Backer wanted only the better boys, not the roughnecks.

Yet Erickson's in Portland stood alone, and was world-renowned for its tremendous capabilities. Loggers and sailors everywhere knew and loved Erickson's. Its sprawl thrilled the sawdust savage down to his wool socks. For many decades from the 1880s, this social center of the nickel beer and the free lunch was the hub of activity on the largest skid road of them all, near the waterfront, and in sharp contrast to the Yankee conservatism for which Portland was widely known. And into the 1970s Erickson's was still operating, although a mere shadow of its former self. Only a small piece remains of "the longest bar in the world" where 1,000 men could belly up to be served by 50 robust bartenders. The great bar, five combined into one, was over two football fields in length, 684 linear feet stretching around an entire block.

August Erikson, host to all the world's floaters, made it big but died a pauper. "The Working Man's Club" was Erickson's own creation. He saw the value of allowing the drinking establishment to grow along the contours of the big land, the big men, and the robust industry it represented. Timber beasts and salts off the steam schooners and windjammers were hypnotized by its overwhelming magnitude. When at a loss describing something large, they'd say, "Why, it's bigger than Erickson's bar," and everyone who was anyone knew what that meant.

Everything but the prices was on that same giant scale, from the massive platters of hefty sausage to the grand pipe organ and the 300-pound bouncers who were, along with the girls, tied into the crimp trade and sometimes confused unruly customers by throwing them out one of many doors, then doing it again as the ousted rowdy staggered around the corner and in a second entrance, seeking another saloon. Five cents brought a healthy schooner of beer, while a quarter purchased two pours of stouter stuff. Mirrors and other fixtures were the finest money could buy, there were no restrictions on calked boots, and the great hall could accommodate hundreds of burly men without necessary elbow-jamming. There was a small stage, while on the mezzanine were cubicles for private entertaining and love-making if a fellow's yearnings went in that direction. And then there was the fabulous free lunch, which was something to behold: roast ox (perhaps bulls that had seen better days on the real skid roads), soft bread cut 1½ inches thick for sandwiches, fist-size slices of sausage, meat and fishballs, soups, broth, stew, steamed clams, Swedish hardtack, pickled herring, and a mustard pot that was anything but dainty. But at Christmas, Erickson's took on real class, serving Tom and Jerry from a punchbowl the size of a bathtub and even genuine manhattans, complete with maraschino cherry.

More logging, it was said, was done within those walls than in all the Northwest woods since logging began. While the timber beast drank and ate his fill, he could contemplate the human misery of a monstrous painting, "The Slave Market," which covered most of one wall. It depicted a Roman slave auction, and in present-day terms, the loggers could "identify with it." They considered themselves slaves to a heartless industry and applied the name Slave Market to their own employment centers, which told of available jobs in signs behind unwashed windows: "one mucker, one bucker, and one PF man."

The lumberjacks' Boswell, Stewart Holbrook, who made a career of chronicling the industry, waxed lyrically about Erickson's establishment and would in the mid-century take eastern guests to visit the fragmentary remains of the old place. Often he was requested to do so, as with his good friends Lucius Beebe and Bernard DeVoto, the western historians who were most eager to view the hangout. Holbrook and DeVoto were photographed in their derbies against the remnants of the world's longest bar, being served by a bartender with hair parted in the middle. It was a scene out of the glory days of Erickson's, but no Portland newspaper would publish the picture, conservative editors horrified to find two of the nation's most famous men of letters—the town's own Mr. Holbrook and the Pulitzer Prize-winning DeVoto—prowling the joints of skid road. The picture finally saw light of day, much to Holbrook's amusement, in Lucius Beebe's flamboyant *Territorial Enterprise,* the Virginia City, Nevada, newspaper which had a national circulation.

The Workingman's Club was normally the prime destination of most timber beasts on the prowl on skid road, but they were often waylaid by innumerable stations that had their own special attractions. Three other key saloons were Fritz', Blaziers', and the House of All Nations, which had pool, billiard, and card tables behind the barroom, and boasted white, brown, and black girls in the upstairs cribs. Mary Cook, Nancy Boggs, and Liverpool Liz were among the more notorious of the "lesser sex" operating saloons. Having long ago liberated themselves from male domination by holding their own with the waterfront toughs, there was nothing "lesser" about them. Mary Cook topped the scales at over 300 solid pounds, smoked black cigars, and did her own bouncing at her joint called The Ivy Green. Nancy Boggs operated a frisky bawdy house informally called a "floating hell hole," which caused the Portland Police Department, bent for a time on holy reform, no little difficulty. Madam Boggs merely hoisted anchor to hover offshore whenever they came around, or if the pressure were too great, headed for the other bank, tying up at the towns of East Portland and Albina. The Boggs establishment had a barroom, dance floor, and rooms above, where some 25 girls resided in lively fashion. The whereabouts of the attractive Nancy's unique houseboat were never certain, as she shifted according to steamer schedules and entanglement with lawmen on both sides of the river, since she refused to pay a license fee in any town. Loggers and sailors had to be on their toes to locate her, but the attractions were alluring enough that even if you had to hire a rowboat, Nancy's joint was always active, somewhere in the port.

A day of reckoning came to pass, however. Officials of the towns grew weary of Nancy's eluding the license fees and were determined to close her down. A simultaneous raid was timed on Nancy's indelicate scow, hitting her from one shore, then the other. Nancy considered this damned unsporting. Grabbing a hose, she poured boiling water and live steam on the raiders, but the hawser was cut, so that Nancy found her scow spinning out of control on the Willamette River's spring runoff, and sure as hell headed to the Columbia and the open sea. Her frightened girls were diving overboard in various stages of nudity while loggers on the docks slapped their thighs and declared this by far the best stage show ever to hit Portland town. But it was unfunny to Nancy. Facing ruination, she leaped into a rowboat and pulled angrily for shore.

Nancy conned a steamboat captain into firing up his sternwheeler to chase the runaway barge. He rushed into action, feeling it would be a calamity if Boggs went permanently out of business. They caught the scow some 10 miles downstream, not far from the mouth, and got a line aboard. After a few strenuous hours, Nancy was anchored again safely in midstream and back in the trade, while the heroic steamboat man had the full run of the place.

Liverpool Liz, who was squatty, weighed 200 pounds, wore a heavy necklace of diamonds and gold, and operated the Senate Saloon with similar gusto. Loggers didn't fear getting rolled in Liz's place (very often, anyway), for she had her own system of reducing their payloads. She could quickly spot a lumberjack fresh from the brush, eager to see the elephant, pockets bulging with six months' wages, enough to choke a bull team. He'd swagger to the bar, and viewing only a few fellows in the room, generously order drinks for the house. At that instant, the bartender pushed a buzzer beneath the mahogany. Down the stairs suddenly traipsed a score of damsels giggling, waving like old friends and calling "Me, too." What appeared to be a cheap round left the lumberjack's bankroll considerably shrunken.

While Erickson's was the largest drinking establishment in the logging world, just down the street stood the Paris House, the biggest damned fancy house in the Northwest woods. Portland lay claim to them both, an astounding fact for the sedate New England city of the West. The three-story Paris House occupied a full block along Davis, housing upward to 100 girls in cribs on the second and third floors. White girls on the second, other shades above, and you pays your money and you takes your choice. In one police raid, 83 strumpets were taken into custody. As with Erickson's, the size of the Paris House alone appealed to the timber beasts, and despite tales of horrifying orgies that went on there, it survived many a reform movement until finally being closed down in 1907, two years after Portland's extravaganza, the Lewis and Clark World's Fair.

Would-be reformers mostly looked the other way, and the people of high society in the lumber towns pulled down their shades. They were influenced by the forces of the lumber industry, deriving much of their income from this freewheeling green bonanza. The skid roads were isolated sections of town; the loggers, seamen, and other working stiffs were considered illiterate bums who lived in a world apart from that of society of the polished dinnerware and silken gowns. What went on along the lawless skid roads was a part of the industry, and even the women had to be rugged to survive. Simon Benson, one of the great philanthropists of the industry, was among the few lumbermen to worry about the conditions of the skid road, largely because celebrating lumberjacks cost him money in lost efficiency and accidents. Benson watched thirsty loggers enter a saloon for beer or whiskey, which was all that could be obtained there, when they might otherwise settle for a drink of water. He therefore gave the city of Portland $10,000 to install 20 bronze drinking fountains in the downtown core area. They are still in use today, one of the city's unique landmarks, with a replica presented to Portland's sister city of Sapporo, Japan. Benson claimed the fountains "helped knock the profit out of the saloon business," but they truthfully didn't seem to curtail the loggers' love of something stouter.

Boss loggers discouraged camp followers in the woods, since they disrupted the work with gambling, drinking, and carousing. The girls who waited tables and cooked in the camps were another kind. The loggers respected them. Besides, after a day of wrestling timber, the loggers had little energy left to wrestle a dame around a bunk. But you couldn't keep the strumpets from trying to get first crack at the payroll, before the timber beasts took off for town. On Vancouver Island, British Columbia, a contingent of harlots set up a house down the road apiece from a logging camp that was off limits to them. To meet the fallers, buckers, and riggers, they took to selling magazine subscriptions as their legitimate business. You had to sign up to be invited to a house party; and soon the camp was filled with lumberjacks subscribing to a wide variety of periodicals, including *The Ladies' Home Journal* and *The Woman's Home Companion*.

Box houses never flourished in the other skid road districts as they did in Seattle, where they were an institution spreading later to the Klondike. The box houses were combined theater, saloon, dance hall, and fancy house, where agile bartenders with fast talk, shifty-eyed waiters, and frisky dames combined their talents at fleecing the customers of every last dime in their possession. The girls who had little or no talent "entertained" with songs and sloppy hoofing on stage, then mingled with the trade, enticing them to buy drinks and food at sky-high prices. The next step was to maneuver a sucker into one of the balcony's private boxes where the dame, teamed with her waiter, could take the entire poke, rolling the unsuspecting male if necessary. The girls were paid in chips for each sold drink, redeemable at the end of the evening. By the time gold was discovered in the Yukon, the Seattle box houses were going strong and were carried to Dawson, laid out by another sawmill man, Joseph LaDue, who saw the land go for $.50 a square foot during the big rush. By then, too, dames like the famed Klondike Kate Rockwell and her boy friend, bartender-waiter Alexander Pantages, knew their trade well, turning it into fortunes in the Frozen North. And it was the skid road box houses that spawned the great theatrical careers of Pantages and John Considine, the vaudeville giants of the Roaring Twenties.

If a lumberjack got rolled and woke merely with a sore head, he considered himself lucky. Shanghai artists worked all the seaports and river towns. The lumber ports were brawling places of call for hundreds of ships a year. Portland by

1880 had 57 shipping lines, the waterfront a forest of tall masts hauling off lumber cut from dozens of mills strung along the riverfront. Many skid road districts were built helter-skelter on planks and piling above the water and with no basements, only the river or bay beneath. A handy trap door was the quick and easy way to rid the place of an unruly customer or to fill out a ship captain's crew, for the captain would pay from $50 to $150 a head. The Bucket of Blood Saloon in Everett squatted above the Snohomish River. There was a spring trap in the back room through which many a bucker and faller who hadn't paid his bill received a permanent dunking, his body later found floating downstream. Or he was given a long sea voyage around the world to think things over. Thousands of lumberjacks were without families or close friends, or were working under other names. When a jack didn't return to camp, little was thought of it beyond the cursing of the boss logger for being short a man. It was assumed that the fellow moved to another camp, even though he may have left his change of socks. No one bothered to check up, not knowing his full name anyway.

The shanghai racket flourished because it was part of the lumber industry system. Ships were sometimes held in port for weeks or months, waiting a cargo. Skippers found it better to let the crew go, then hire another, than keeping them on the payroll. It was far cheaper paying blood money to the man-snatching pirates or crimps who sometimes even boarded incoming ships to kidnap a crew. Blackjacks, dope, dance hall strumpets, booze, and the forerunner of the Mickey Finn were all employed to get a crew together. The crimps didn't care about experience, and neither did the tough skippers who found themselves dealing with two or three experienced mariners and 30 or 40 loggers and plowboys. The lumbermen needed to get their product to market; it did no good resting on a dock. A windjammer or steam schooner riding dead at anchor for lack of a crew was time wasted, dollars lost. And the price of lumber was so flexible that it might go down at any moment. Little was done to stop the crimping along lawless skid road, save by the loggers and sawmill hands themselves. The system kept the cash flowing; it was profit for the lumberman, profit for the saloonkeepers, profit for the crimps. The odds favored the latter, for while a lumberjack might rant and rave with anger upon awakening aboard a creaking tall-master far at sea, vowing to "get that sonofabitch," by the time he reached home six months to a year later, he'd cooled down considerably. One timber beast even declared gratefully that six months on the bounding main cured his consumption.

Among the shanghai ports, Portland and Grays Harbor were the most notorious. There were several notables—Jim Turk, the Grant brothers, and Larry Sullivan—who operated Portland "boarding houses" where sailors, loggers, and other luckless stiffs were convinced one way or another to see the world. Joseph "Bunco" Kelly grew into a lyrical legend, the most wily of them all. What Kelly got away with along the Portland waterfront is more than slightly amazing. It was Bunco who once filled out a ship's crew with a cigar store Indian, wrapping it in a tarpaulin, and lugging it aboard, collecting his $50, and explaining to the skipper that the big fellow was plenty liquored but would make a fine hand when he sobered up.

Kelly was forever the opportunist; it is a wonder some logger or ship's mate didn't bash in his head. In one situation, rowdy loggers on a spree broke into a basement near Snug Harbor, thinking it the storage room for the bistro. In truth, it was a mortuary. There were barrels all around, a bountiful bonanza, and the jacks hove to, downing a keg containing formaldehyde. Soon the mortician held more business than he would normally receive in a month-plus, with 39 dead men stretched out on his basement floor. But while the mortician slumbered, Bunco Kelly was ever alert. He'd been trailing the hard-drinking loggers, since he had a large order to fill. He viewed this wonderful windfall through the basement window. Hiring every rig in sight, Kelly hauled the corpses dockside and onto the outgoing vessel, just about to sail, and managed to shake down the happy skipper an extra $2.00 a head, since "I got these guys drunk on me own money."

For many years Kelly haunted the waterfront dives. There's no telling how many hundreds of unwanted passports he issued. Eventually changing times caught up with Bunco, though. He was sentenced to the Oregon Penitentiary, where he was able to identify the body of outlaw Harry Tracy following the killer's last rampage through the Northwest, and he even wrote a small book about his experiences there.

The crimp trade went on for many years in Portland, and there are men who remember it well. An old-time sternwheeler captain, Sidney J. Harris, who guided the steamboat *Henderson* during the making of the motion picture *Bend of the River* in the Northwest in the 1950s, recalled looking down upon a crimping scene in broad daylight from a second-story vantage point. Several men ganged up on a victim, who put up a valiant fight but lost, and "Happy" Harris was told it was a shanghai. Some of the old skid road buildings, mostly succumbing to urban renewal, have long since nailed up their trap doors, but hidden stairways beneath rustic old docks are still visible.

The pattern was the same from Eureka to Vancouver, British Columbia. Port Townsend was plenty dangerous. As port of entry to Puget Sound, Townsend was a natural for the crimping trade. The dives were strung out near the beach, each equipped with its sets of trap doors and deadfalls. It was a simple matter to make the drop after the victim was pulverized with knockout drinks; then the limp form could be lugged unobserved along the beach to the waiting ship. Limey Dirk's Sailors' Boarding House (these places alone were a tipoff for all but the greenhorn and unwary) and the Pacific Hotel were good places to be given wide clearance. Still, the lumberjacks and sailors seemed to enjoy this game of fox and hounds. It was part of the only life they knew, and some of the so-called "best people in town" engaged in the racket, among them the Norwegian vice consul at Aberdeen. Only Indians working in the woods and mills were safe from the shanghai snares. The crimps feared that kidnaping Indians might bring the almighty wrath of Uncle Sam down upon them.

Seattle's skid road was the original, Portland's the largest, and Aberdeen's the toughest—very tough indeed. Aberdeen, also called Plank Town, was branded among the wickedest cities in the country, and it was a long time before it lived down that reputation. Plank Town was the Port of Missing Men, "where they picked up a body bobbing amid the log booms most every morning." It was wild beyond imagination, more dangerous than New York's Bowery and San Francisco's Barbary Coast, uncivilized and bawdy, with its motley waterfront collection of saloons and bistros making a western mining camp appear as tame as a church social. The booming lumber port, population 10,000 in 1908, and its sister cities of Hoquiam and Cosmopolis were wide open.

In the years of the Big Bonanza, thousands of lumberjacks, sawmill working stiffs, and seamen tromped its lower planked streets seeking escape from boredom in the rustic booze parlors. George the Greek never parked his lunch wagon before a saloon door, lest a customer be given the heave-ho. There were 50 or more saloons, all told, and a fellow could acquire quite a hangover attempting to test the quality of the booze in all of them. But many a jack gave it the old college try until the names blurred and the joints ran together almost as one—My House, Log Cabin, the Grand, the Humboldt, Klondike, Whale, Book, Mecca, Lion, the Palm, Harbor, Board of Trade, Crescent, Pioneer, Fairmont, Star, Fashion, Washington, North Pole, Alaska, Moonshine, California, Central, Walker, Royal, Combination, and Mint and Loop, to mention the most notorious.

The town had a good thing going, licensing the saloons and collecting tribute from the fancy houses, including madams and procurers, without openly endorsing them. On a June day in 1907 100 girls were "arrested," taken to police headquarters and fined $10 each as inmates of the houses. Procurers were assessed $25 apiece. This became a common tactic whenever the city coffers were depleted. The Harbor's saloons, blossoming from eight in 1895, required a $1,000 yearly license fee. Aberdeen alone derived over $20,000 annually from the saloon business, one of three prime sources of revenue. Taxes brought $25,000 and the water service, $18,000. When Washington State had the audacity to vote prohibition, it shook the lumber capital to its corks, and the license bureau clerk scrawled across his final entry, "Last of the Wets." It was an upsetting time indeed.

The Grays Harbor skid road ran at full blast 'round the clock, and the revelry could be heard across town and bay. Yet a logger packing a considerable load ran more than the normal risk of shanghai because of Billy Gohl, especially if he stashed his poke at the Sailors' Union Hall above the Grand Saloon at Heron and F streets. It was later discovered to be the most dangerous trap in town, and in the end it brought the law into skid road.

In the late 1960s and early 1970s, a number of shocking murders occurred in this country. Multiple killings became commonplace. Among them was the gunning down of students by a berserk young man on a Texas university campus and the murder of eight student nurses in Chicago. Then in 1971 the bodies of some 25 victims were uncovered in an orchard near Yuba City, California, leading to the arrest of a farm boss who was allegedly doing them in for money. Newspapers and wire services, dealing in characteristic superlatives of those years, branded the affair as "the greatest mass murder of all time." Yet the Yuba City case failed to come close to the bloody work of one William Gohl of Aberdeen, whose very name fit the situation.

Billy Gohl came to Grays Harbor following the Klondike Stampede, where he may have worked for the infamous Soapy Smith at Skagway. Gohl was stocky, with thick shoulders and a heavy chest, a bull-like neck, and the appearance of a rugged individualist who could more than hold his own against The Harbor's tough element. Billy was personable on the surface but downright mean beneath. He parted his hair in the middle, bartender-fashion, and he had flashing blue eyes and a thick mustache, which gave him an appearance of honesty and good humor. Yet behind those eyes and happy chatter lived the most ruthless killer ever to walk the boards of an American seaport, or anywhere in the western frontier.

Gohl operated beyond the farthest extremities of the law and was quite open about it, boasting of his deeds. He was bound up in the labor movement and the Wobblies. He burned hotels, sawmills, houses and businesses, burgled homes, pirated ships, and shot down men in cold blood. Yet Billy posed as a friend of all the lumberjacks and sailors of the waterfront, especially the lonely ones off a ship or from the woods who were looking for the elephant. He could spot them quickly and offer a newcomer friendly advice on how to get along in Aberdeen. He lectured the boys about the flaming dangers of this mighty hellhole, suggesting that it would be wise to leave their pay with him for safekeeping, rather than be rolled while on the town. Certainly as agent for the first Sailors' Union on the Pacific Coast, he was trustworthy enough, the good friend of the working stiff.

Gohl's boldness was astounding. Once his gang boarded the vessel *Fearless* to force off the crew, as he had done numerous times to urge the hiring of union men. This time the skipper stood his ground, bellowing that his crew was satisfied. Guns exploded in a battle that lasted an hour, and Gohl's forces were driven off. The captain filed charges, and Gohl paid a $1,200 fine, unwilling to take things any farther, as the federal agents might come snooping around.

Billy had good reason to keep clear of the federal men, since he'd been involved in so many acts of violence in the Grays Harbor country. He had many enemies; the lumbermen, ship owners, skippers, and businessmen had no use for him. If the opportunity arose to even the score, it was quickly seized upon. As a symbol of his cockiness, Gohl had painted the privy hanging over the rear of his union hall a brilliant red. Heading down the Wishkah under tow, skipper Michael McCarron of the *Sophia Christenson* couldn't resist. He swung the wheel so the ship's long jib boom clipped the outhouse as neatly as topping a tall fir. Downstream went the schooner with the two-holer swinging jauntily from the broomstick, while Gohl cursed and shook his fists in anger. Too bad, muttered McCarron, that he wasn't inside.

For a long while there had been the regular disappearance of loggers and sailors without apparent reason. It was generally passed off that they'd been shanghaied or went to work down on the Columbia or Coos Bay shows, in the redwoods, or over in the pine country. But when the body of a friendly logger, thought to have left for another job, was picked from the Wishkah late in 1909, head bashed in, the lumber community was visibly shaken. It was one thing to have transient sailors killed, another to do in a lumberjack. The situation worsened; through the winter some 43 bodies were found in the log booms or floating on the outgoing tide.

Aberdeen business interests, dependent on the industry, met in secret session, as had the western vigilantes of yesteryear. They put up $10,000 to bring the killings to an end and apprehend the murderer. The committee hired George Dean, constable at nearby Cosmopolis, who was given a free hand and carried simultaneously 14 sets of coroners' juries to sit at inquests over the bloated corpses picked up in the rivers.

Dean brought into Aberdeen a pair of Thiel Agency detectives using the names Paddy McHugh and Billie Montana, who soon purchased the saloon beneath Gohl's office with some of that $10,000. Gohl was a prime suspect because of his violent union activities during this time of labor crisis with the IWW at The Harbor, and because he could readily identify many of the sailors' bodies as men who visited his union hall. Dean realized that Gohl was closely tied to waterfront life, but when he learned that many of the victims had stashed their pokes with the agent, his suspicions grew stronger.

The long trail of evidence and the statements of a secret witness led to Gohl's establishment. The horror of many years began to unravel, including fires, bombings, and killings far above the 43 of that winter. Gohl's brutality and cruelty were difficult to accept, even in wild Aberdeen. Once Billy encountered a Swede who asked Gohl to hold his poke while he made the rounds of the bawdy houses. Gohl got him to dress as a logger and go sit on some pilings outside the union hall to "watch for a scab boat." Then he shot the fellow through a window with his rifle and saw him tumble into the water. At another time Billy coldly abandoned several "scabs" on a small sandbar at low tide. None could swim, and the rising tide took them all. In a third instance, he held a gun on several pleading victims, watching them drown as the tide came up around them.

The total of known victims ran over 100—124, according to one count—and may have been well over 200, for in his first years, Gohl used a launch, transporting his victims across the bar to a deep-water grave in the Pacific. Then he grew lazy or careless or both, cocky with self-confidence, and this became his undoing, for he dumped the corpses into the rivers or buried them in the mudflats. It became apparent that "Murder, Incorporated," as he was called, had been operating his one-man den of horrors for years. In later and calmer times, crews laying a modern sewer system for the city dug into ground filled with human skeletons. The workers would comment grimly, "Another of Gohl's catacombs . . ."

Billy's trial was a sensation. Copies of the *Aberdeen Daily World* were snatched up before the ink was hardly dry, and the courtroom couldn't handle the onlookers. He was sentenced to life in the Washington State Penitentiary at Walla Walla. However, there was a strange denouement. Years later, Gohl broke completely at the sight of a desperate black man driving a knife into another prisoner in the mess line, believing him a stool pigeon. Gohl collapsed and was transferred to the "asylum for the dangerously insane" at Sedro Woolley, Washington, where the "Mad Monster of Aberdeen" died in 1928.

1. The thick, virgin forests of the Pacific Northwest were considered a wasteland by the early settlers, who burned them to clear land for farming. They called it all the Green Desert. —*Oregon Historical Society*. (Chapter II)

2. John Porter attempted to outdo even the big trees of the Oregon country by bringing back sequoia cones instead of nuggets from the California gold fields. Porter planted this magnificent avenue, which stands today near his pioneer home. —*Ellis Lucia photo*. (Chapter I)

3. When in 1948 a homeowner ordered the ax to twin sequoias that towered above all else at Forest Grove, Oregon, it created an ecological crisis years before the word was in common use. —*Author's collection*. (Chapter I)

4. Port Gamble was a transplanted New England village, patterned on East Machias, Maine, with a dignity and style unlike most of the Northwest's roaring logging camps. —*Pope & Talbot.* (Chapter II)

5. Musclepower and brawn with sharp ax and huge crosscut saws called "misery whips" conquered the giants of the Big Woods. Often loggers did the impossible, applying ingenuity and stubbornness in getting the timber down to splash. —*Washington State Historical Society.*

6. Oxen dragged the great turns of logs to landings and sawmills over what became the skid roads. —*Oregon Historical Society*. (Chapter II)

7. Rustic logging camps deep in the woods were different from those of the West's mining centers. The camps moved with the timber falling. Note bull teams at left and horses at right, the only "power" early loggers possessed. —*Oregon Historical Society*.

8. Logging crews were young, full of vigor, and usually large in size. The work was tough, the hours were long. Gradually women invaded the deep woods as cookhouse crews. —*Author's collection.*

9. Falling and bucking these giants with crude tools were no simple tasks. Loggers worked on springboards notched into tree trunks. There was much waste making the first cut high above the ground. —*Oregon Historical Society.*

10. Small woods mills, operated at first by water and then by steam, rough-cut lumber for local markets. —*Oregon Historical Society.*

11. The sawmills grew larger and so did the lumber fleet. This scene is likely Bellingham, Washington, on Puget Sound. —*Oregon Historical Society.*

12. The skid road districts (never skid row) of the lumber ports were the loggers' domain when on vacation or between jobs. Erickson's Saloon in Portland contained the "longest bar in the world," famous around the globe. This is the card room and only a small part of Erickson's, which still exists in Portland. —*Oregon Historical Society.* (Chapter IV)

13. The easiest way to move logs to the sawmills was by water. The Big Woods had thousands of natural arteries. This is the Portland harbor, with six-master in center background and sternwheeler at far left. Logs are still moved in this manner today. —*Oregon Historical Society*.

14. With tender, loving care, loggers built the world's biggest log cabin for the 1905 Lewis and Clark Centennial Exposition, in Portland, from huge, choice timber. It became a landmark, lasting over 50 years. —*Ellis Lucia photo.* (Chapter VI)

15. The avenue of trees of the Forestry Building was an awesome sight and a tremendous feat for its time. Decades later, visitors still wondered at its size and construction. By then the building couldn't have been duplicated, for the big old growth was no longer available. —*Oregon Historical Society*. (Chapter VI)

Chapter V

The Big Cut

If the hell-roarers are logging the hogbacks beyond Valhalla, and surely they must, it would be in a stand of Douglas fir. They wouldn't touch anything less, or find anything better. The straight, tall and great-girthed timber, sweet as the waftings of heaven itself, would be growing on endless hills, section upon section as far as the eye could see. And the old bull whackers would be hammering the canyons with mighty oaths, and down the skid roads would stomp the hairy-eared giant-killers with chips in their gizzards and pancakes in their craws. Their boots would be spiked, their pants staggered, and their wool underwear would rasp an ordinary mortal raw. They would be the stompingest, cussingest, ring-tailed hellbenders ever to pass from one paradise into another. It would be just like Grays Harbor when it was young and the sap was running strong. . . .

ED VAN SYCKLE
Aberdeen Daily World
1963 SPECIAL EDITION

Twelve thousand lumberjacks flocked to the Big Bonanza of the Grays Harbor country from all over the world to take part in mining the richest, the finest, the most beautiful timber stand on the face of the globe, which extended ever yonder into the stream-slotted mountains. It was indeed the Forest Primeval, the mother lode where for unknown centuries the Pacific's ravaging storms had deluged the land with incessant rains—the lifeblood of this healthiest of all timber.

Grays Harbor was the last stand of old-time Paul Bunyan lumberjacking, beyond the wildest hallucinations of a well-oiled logger after the Fourth of July blow-in. Two million breathtaking acres contained billions of board feet of virgin fir, spruce, cedar, and hemlock of butts 12 feet in diameter, stretched thick as grass to hell and back, reaching for the high clouds, alive when the Vikings came to the continent and well on the way to giant status when that latecomer Columbus arrived. Strange Chinese junks had floated across the Pacific, landing on the broad beaches and secluded bay tidelands of the Washington and Oregon shore; and there was a Russian colony near here for a brief time in 1808, a few seasons after Lewis and Clark floated down the Columbia to winter 100 miles south of Grays Harbor. Many had seen these grand trees, but few found anything to become excited about.

Word of the Big Bonanza spread like a forest fire to the gutted timber belts of the Midwest and into New England. The lumbermen yanked down their stacks, uprooted the boilers, rolled up the belts, bullchains, and band saws, bound up the big circular blades, and shipped the whole shebang to Grays Harbor to make war on those mighty timber stands. And the lumberjacks streamed west from Europe—rangy, rugged Norwegians, Swedes, and Danes— whose people had cut timber for generations in that part of the world, in country not far different from this, and unlike the wasteful and less visionary Americans had long ago learned that whatever you took, you had to give back, for the supply wasn't endless. In their own land they had developed wise and proven methods of forestry to perpetuate that supply. But the Americans were, as always, in a hurry, and why, hell, the timber around Grays Harbor was so thick that you'd never run out. It would last for generations. They sank their double-bladed axes and great saws in trees so rich with life that the pitch spurted like blood from a new wound; and this grandest timber creaked and groaned in agony as it swayed in shock and one by one went crashing to the ground. The fallen giants were sliced into sawlog length, and weighing upward to 40 tons, they were sent tumbling on their way to the mills, which sprang up like mushrooms along the bay and deep-running rivers.

Everything was ripe for the cutting. Conditions were ideal. If God didn't wish this wonderland harvested, reasoned the lumberman, he wouldn't have grown it this way and made the timing and the layout of the land so perfect. The many streams were natural canals to carry the huge sticks down to the bay, where the sawmills sliced them into boards for a world market. That marvelous bay beckoned the lumber fleet—hundreds of vessels, and well over 1,000 in a single year—across the dangerous bar to slam their broad bellies and shove their sharp prows against the mill docks of Plank Island; then outward bound, hauling billions of lumber feet to San Francisco, New York, Boston, Cape Town, Kobe, Shanghai, Sidney, the Hawaiian Islands, Batavia, Antwerp, Southampton, and hundreds of other ports of the world.

Things were slow-paced at first, despite the eagerness to fell the trees. The gear was crude, most of the power coming from the muscle of lumberjacks and their bull teams and from the fast-running streams. A few early arrivals, mostly farmers, were jackrolling logs when the first bulls arrived, but this was only the handy stuff that you could prod into the water. The pioneers found themselves long on four-letter words and short on motive power, for the sticks were too damned heavy to handle or haul very far. The professional logger was therefore a welcome sight, with his springboards, calked boots, misery harp, long-handled ax, frayed tinpants, and red underwear and hat, which were the marks of his proud trade. The pros could probe deeper into those timber veins by using the skid road method. The skid road builders brought in their four to six spans of oxen to lean into the log turns, moving to the squeals of the rides, little wisps of smoke rising from the skids, and the standing timber shaking to the oaths administered by the stout vocal chords of the bullwhackers.

The skid roads were prophetic of things to come around Grays Harbor. As these roads penetrated deeper and deeper into the forests, the lumberjacks were able to invade new stands, which were needed, since the handy good sticks were already running out. They were felling only the big stuff, the tallest and best, leaving what smaller timber could survive the onslaught of crude logging techniques, which gouged everything in their pathway. In a way this was selective cutting, which would become a matter of heated debate between lumbermen and ecologists a generation later, but the selection was done from greed and for dollars rather than with anything in mind that came close to being called conservation. To their mind, it was nature's problem to heal the wounds.

In the late 1880s, Cy Blackwell, a state of Mainer, brought in the first Dolbeer donkey. Although it took a little time, Grays Harbor began shifting gears toward highball to make up for the lost years; and if ever the cut-and-git philosophy abounded, it did so around The Harbor, for there wasn't a logger or lumberman in 10,000 who cared about what was happening to the forests.

The size of those trees set the scale for everything. Much of the industry's heavy logging and sawmill gear was manufactured, of necessity, right at Grays Harbor, for eastern machine plants couldn't envision what was demanded. The spinoffs from this industry were as numerous as sparks flying off a whetstone. There would be a town, or maybe several. Samuel Benn, who was Irish-born and a farmer, saw the great potential in 1888 for his bottom land as a mighty city, deciding to call it Aberdeen. Benn, who lived to be 103, long guided the destinies of his town. He had the satisfaction of seeing it and neighboring Hoquiam, meaning "hungry for wood," reach a population of 40,000 to become the industrial giants of the Pacific Northwest coast by the 1906 San Francisco earthquake.

Aberdeen was going great guns in 1903 when it caught fire, much of the core area of ten blocks burning to the ground. The fire started at the Arctic Hotel in the Mack Building on Hume Street, killed four people, and destroyed 140 buildings, many of them uninsured because they were "flimsily built of inflammable material"—lumber! A few brick buildings survived. For nine hours the holocaust raged before being brought under control by using dynamite supplied by Dr. Paul Smits, a local physician. How he used dynamite in his practice wasn't explained. But the explosive couldn't save the fine Aberdeen Opera House, a point of pride and local culture in contrast to the rowdy ways of Hume Street, at a loss of $20,000. For a brief time, too, the saloons were closed to prevent looting, something of a major frustration to the community, which existed on booze and the free lunch.

The town was quickly on the rise again, for there was plenty of material for rebuilding, and fires even as terrifying as this one were a way of life around Grays Harbor.

"The city will gain by the fire," commented founder Sam Benn optimistically. And gain it did. The San Francisco quake three years later provided the initial impetus to the unbridled boom times that extended through World War I and the Roaring Twenties, lasting into the Great Depression, then reviving for

World War II. The unprecedented demand for lumber to reconstruct the earthquake and fire-stricken city by the Golden Gate was opportunity knocking at The Harbor's door. Unlike the early pioneers who didn't have the mills to meet the lumber demand of the Forty-niners, Aberdeen could handle it. There were at least 15 mills along the waterfront; and now with steam donkeys and railroads coming in, the great logs were moving faster and faster from the hills to feed the hungry plants. The sawmills could manufacture 100,000,000 board feet annually, the glow of their waste fires lighting the night, smoke and sparks belching into the moisture-laden sky, and the towns so boisterous that when the shifts changed, the thunderous roar from thousands of corks along the boardwalks of Plank Island sounded like a cattle stampede.

Mechanization in forests and mills opened the Big Woods, and the invasion by the railroads on the heels of the steam donkey created the gol-derndest network of flimsy trackage ever imagined by civilized man. But before the iron horse and for a long while thereafter, the river log drives reached fantastic proportions. The drenching rains allowed logging outfits to depend upon swollen rivers like the Humptulips, Wishkah, and Willapa for getting the timber to the marketplace.

"We have June freshets, but they don't always come in the same month," pointed out an old-timer.

You couldn't count on the timing of semi-annual flooding, and the sawmills must be fed. Grays Harbor demanded inventiveness to meet the needs of this unpredictable land, so someone devised the "splash dam." The splash dams did the job, creating their own log runs. The dams were operated by the boom companies, which worked the rivers as common carriers, although some of the bigger logging outfits had them, too. Built in series, they were beyond the imagination and understanding of the average woods logger.

"The dams themselves were hellers to build," wrote Ed Van Syckle in his colorful style in the *Aberdeen Daily World*. "They were made of logs, cribbed up, shored, braced, cross-tied, driftbolted, and bound with wire line. They had multiple gates and sluiceways so that a full head of water could be released at once. It took a crackerjack of a builder to get all the pieces in the right place— men like Jack Byard, who built more than 30 such dams without ever having one 'blow out'; or Bob Turner, a masterfully cussing hothead who in a fit of temper would pick up a logging jack or some similar trinket and heave it into the canyon.

"Next to the molly hogan splice, or maybe a Cape Horn button or springpoles for the donkey whistle, the false-gate dam was somewhere near the top in ingenuity. This cumbersome but ingenious device permitted a small cascade of water from a lifted splash board to strike a false gate hinged at the bottom. Down went the false gate under the weight of water, and up popped the main gate, the two being connected by cables through overhead sheaves. Things began to happen. The gravel bars below were swallowed up in a surge of water bobbing with logs. Crows lifted from the riverbed, squawking. Ranchers came down from their barns and cussed loudly in innumerable tongues and accents. The blue haze that frequently hung in the valleys was attributed in part to the scorching things that were said of the men who owned the booms and dams.

"Every sluicing or splash was a gamble whether a rancher would lose 10 feet of meadow, or all of it. The water swooshed against the deep-soil banks, logs came hell-for-leather and gouged out great chunks of fine loam, which would be noticed days later off the Washington coast as a discoloration of the water. Some horny-handed sailor would peer overside and tell himself the ship was nearing Grays Harbor.

"The ranchers were no pale-version pioneers. They were rough and rugged, and could talk a logger's language any day of the week, with embellishments. And they didn't cotton to the idea of having their oat patches, or maybe their hen houses, cut away from under them. Individually, they made it pretty tough for the boom companies, but when they banded together and hired lawyers, they often tied up the rivers tighter than a pig's eye in fly time. One individualist down on the Humptulips sued a boom company every year just as regularly as he planted spuds or weaned calves. Just as regularly he collected damages. It became sort of a license fee the company paid for using the river past his property. But one year there wasn't any suit. Still the boom company sent the check anyhow, the bookkeeper thinking there had been an oversight somewhere. It was later discovered the rancher had a fractured leg and couldn't get to court."

The dams were stout affairs, compiled of giant timbers. The Malinowski dam of the Wishkah Boom Company's main river operation was 300 feet long, had six gates, a maximum head of 45 feet, and a 15-foot draw-down. When raised to create the "splash," the six gates would draw down the head eight feet after 90 minutes. The Humptulips Boom & Driving Company, another big outfit, had eight dams on the Humptulips alone and was able to handle driving for several logging companies, figuring the cost on volume. The system got the logs down to splash, but the dams added to the tremendous waste of this sawdust empire, which was already being buried by its own debris. Some of the mighty timbers were just too huge to navigate the splash runs and hung up, high and dry, on the banks. There was no possible way to budge them, save perhaps by a cloudburst, and millions of board feet of raw lumber lay for years along the rivers, too heavy for peavey power and awaiting the time when the gasoline-and-diesel-driven tractors invaded the woods to make salvage possible —and profitable.

The railroads altered the picture, for Grays Harbor went for the iron horse in a big way, although the river runs lasted well into this century. There were twenty millbound logging railroads hauling timber from back in the brush. The "dinkeys" of the back-country operations, using small flatcars and detached trucks, took the peelers out along roller-coaster systems that would give any respectable railroader nightmares. Following the holocausts of 1903, the fabulous Polson Brothers, one of The Harbor's greatest companies, began setting the pace for railroad logging. The volume of fire salvage was so huge, reasoned Alex and Robert Polson, that railroad logging would speed the logs to the mills. The opportunity was there; new plants were rising in the tri-cities of Aberdeen, Hoquiam, and Cosmopolis; and there was a free market for logs. The Polsons were soon in this salvage effort for keeps, for Merrill & Ring had millions of board feet of firekill and gave the Polsons a contract to haul it out.

The Polson Brothers Logging Company was already a big outfit, with a large sawmill and a shingle mill that produced 250,000 shakes a day. Alex

Polson, who came West looking for gold, had operated sawmills and lumber-jacked in many parts of the frontier, but saw his greatest opportunity at Grays Harbor. He sent for his brother, Robert, and the pair formed a pio-neering kind of company that was always trying new ways for getting out the big sticks. In time, they became the largest log producer on the Pacific Coast: They were owners of many sprawling timber stands north of Hoquiam; a leading dealer in logging machinery, supplies, and equipment; boasted one of the biggest logging railroad systems; had their own tug fleet on the bay to haul log booms to the mills; had investments in British Columbia copper and lead mines; and were a powerful influence during the years of the Big Bonanza. Furthermore, the Polsons encouraged their workers to plant flowers at the camps and on cutover lands to lessen the ugliness, which was something of an amazing concession.

Opportunities seemed everywhere in those boom years. Many gambled; some won, some lost, and not everyone had the know-how and physical muscle of the Polsons. Most lumber was still being shipped by water while the lumber-men got the railroad itch. The Harbor became connected with the inland America by iron trails. The Northern Pacific thrust rails to Montesano in 1890 under the banner of the Tacoma, Olympia, & Grays Harbor Railway. The following year, it reached Cosmopolis and the strange boom town of Ocosta before touching Aberdeen. Developers captured the railroad, determined to make Ocosta-by-the-sea on the south shore the leading metropolis of The Harbor. This upset Aberdeen's founder, Sam Benn, and also local business interests of his town and Hoquiam. The hometown citizenry was not to be denied. Their own railroad would be built to the junction with "volunteer labor"; every able-bodied man would give 10 days' labor, or the cash equivalent of $2.00 a day. Lumbermen J. M. Weatherwax, A. J. West, and Charles Wilson, as organizers of this enterprise, donated rails and crossties. The rail was from the British bark *Abercorn,* wrecked in 1888 in a thick fog 10 miles north of The Harbor, with only three survivors. Five years later, Captain George A. Pease, one of Grays Harbor's more colorful personalities, built a long trestle out to the wreck to salvage 2,000 tons of rail. The salvage operation was long and costly, extending over six years. When the rail was finally brought ashore, it was found to be pitted from the salt water. No matter; the iron went into the Plank Town spur. Night passengers of future years knew when they were ap-proaching Aberdeen from the sound the wheels made on those pockmarks. But the Aberdeen branch helped the brawling lumber port maintain its status and puncture the Ocosta boom. Despite its shingle and lumber mills, flour mill, brewery, business buildings, and fine homes, Ocosta faded into oblivion, and the Aberdeen branch became the extension of the main line.

Developers were determined to get their hooks into some of that dough floating so loosely about The Harbor. Three miles west of Hoquiam another "boom town" came into being—on paper, anyway—under the auspicious name of Grays Harbor City. Lots sold for $500 and up. Within three months, all the land was gone. The developers created a flurry of activity. Streets were graded, business buildings and private homes were built, a brick factory was established, and a mile-long wharf was extended into deep water of the bay. The railroad was the impetus behind all this, but when the iron horse failed to come that way, Grays Harbor City's balloon burst in favor of nearby Hoquiam, which gained a $100,000 hotel, a bank, a sawmill, and a sash and door plant.

The Harbor became self-sustaining, with assorted companies and manufacturing plants springing up to handle the needs of the booming industry, from huge tackle blocks to locomotive sprockets and the great saws required to slice the giant timber. There were good years and bad, but the green gold kept coming from the woods, far up the Olympic Peninsula, where the legendary Iron Man, John Huelsdonk, had his place miles from any road. His good wife, nevertheless, had a kitchen range, packed in with ease by her husband, who was rated as the toughest man alive, close as you could get to Paul Bunyan. Some back-country settlers swore they saw Huelsdonk packing two stoves at once, but this was doubted as merely a tall tale, since there weren't any two back-country families in the Hoh Valley who could afford new ranges.

The year 1906 began as a bust, aided by a very dry spring. The San Francisco earthquake had created an instant demand for lumber, and prices skyrocketed, but the lumberjacks couldn't get the logs out of the woods. All the streams were low, so millions of board feet lay in the back country, unable to reach the mills. The lumbermen prayed for rain as they never had before, and hoped for a miracle, a rousing freshet. The good Lord was with them: For one solid week water poured from the skies.

Logs valued at $7.50 a thousand one week were now worth $12 in the log booms. Grays Harbor hit the jackpot and was wading knee deep in wealth. Much of it was dumped into the skid road saloons and the stockings of the harlots. The boom was still going strong in the panic of 1907, but silver was scarce, and the local banks "raided" the saloons for coin to meet the payrolls. The silver discs went spinning in a round robin, right back to the bars and dice tables of Hume Street, where the barkeeps were so busy raking it in that they didn't even bother counting the coin until it was needed; they just dumped it into wooden boxes that would hold $500.

The lumberjacks demanded hard money—gold and silver coin—because it was durable. It was one reason Pope & Talbot adopted its policy of paying in silver dollars from the San Francisco mint. Paper was no good; loggers' pokes sometimes got soaked from an unexpected dunking in a river or mill pond, or maybe when caught in a heavy squall. In the fall of 1907, the nation was struck by a severe shortage of "real" money. Grays Harbor was suddenly unable to pay in either coin or currency, the mills booming and the workers demanding their wages. Trying to explain the situation to several thousand tough lumberjacks and sawmill workers wasn't a happy prospect.

There simply wasn't any money in the midst of these local boom times. Early on November 8, Banker W. H. France of the Montesano State Bank was notified that the First National in Portland was unable to ship any more money to Grays Harbor "until further notice." France recognized it as a crisis that called for drastic measures because of The Harbor's peculiar way of life. He contacted W. J. "Billy" Patterson of the Hayes and Hayes Bank of Aberdeen and W. L. Adams of the Hoquiam First National. The bankers decided to print their own money.

Rushing to Tacoma, they ordered an immediate printing of $300,000 in currency labeled "Certificate of the Associated Banks of Chehalis County." The bankers were in a hurry, for there was payroll to meet. Workers at some plants drew pay daily, blowing it that very night along the skid roads. The men might not like the paper stuff, but if it were explained to them, they'd probably accept it, so long as it bought what they wanted. One thing about loggers: They were ready to accept reasonable improvisations such as this strange money, which bore printing only on one side of each certificate, since there had been no time for any double runs.

The bold bankers gambled on their own feelings about the men of The Harbor, and they were right. There were several more printings, totaling $1,500,-000. Banks within the county were required to post collateral, often customer notes, before being issued any Grays Harbor Money. Superior Court Judge C. W. Hodgdon, who had a long, flowing beard, dispensed the needed funds on approval of France, Patterson, and Adams. The paper, illegal by federal law, filled the great gap in the local hunger for gold and silver coin, keeping the mills humming and the men working. In about three months, the crisis was over and Grays Harbor Money was redeemable at the banks, which had guaranteed its full value. Then as a final act, cashier James Fuller and Judge Hodgdon scooped $1,500,000 in one-sided illegal certificates into a firebox at the Northwest Mill before the disbelieving eyes of the firemen. It didn't quite come to the full million and a half, however, for a few certificates found their secret ways into safes and private boxes, worth much more as "collectors' items" today.

That was how The Harbor did everything, on a glorious scale using the same raw courage as with printing its own money. Whatever the need, there was someone around to meet the demand. The bonanza towns of the First World War and the Roaring Twenties were supported by 37 sawmills plus innumerable spinoff plants, and that is something to reflect upon. A total of 200,000,000 board feet of lumber were manufactured in the peak year of 1925 by 19 tidewater mills, using mammoth 72-foot band saws, while the balance of the year's output—1,563,736,000 feet—was turned out by 18 mills in the area, funneling their product on to Grays Harbor. The mills' payrolls reached 5,900 men, while an additional 4,200 lumberjacks were sweating and cursing back in the wet woods for 10 large logging companies and a collection of gyppo outfits.

This was only part of it. The Harbor produced just about everything imaginable in the way of lumber products—doors at the rate of 680,000 annually, shingles by the millions, veneer, lath, piano sounding boards, bucket food containers, box shooks, pulp for paper, veneer siding for orange boxes, molding, casing base, garage doors, and a variety of miscellaneous cabinet and factory specialties. This meant jobs for another 2,000 men. Manufacturing records were established; the Henry McCleary Timber Company, for one, was turning out 6,000 doors daily. The lumber towns had great machine shops, for with that much manufacturing activity, something was always breaking or wearing out, and the cry was for instant replacement. The lumbermen couldn't await shipments from San Francisco.

Some of the world's biggest logging gear was made right on the scene, not only to wrestle the giants of Olympia but also for shipment to other big timber areas, including the California redwoods. The sprawling Lamb Machine Company plant at Hoquiam, with its six-ton ladles and two-ton electric furnace, purportedly one of three in the country, became legion for its manufacture of massive logging blocks, weighing thousands of pounds, and the development of "lamsteel," specially treated for additional strength and hardness to withstand the rough grappling by hooks and tongs. Lamb's mighty plant was the last word in updated efficiency and included a drafting room, large pattern shop, foundry for casting steel, iron and brass, forge and machine shop, ample rail facilities, and warehouse space for storing a month's supply of its standard products. The mighty logging blocks were the main items of manufacture, but the company also supplied new machinery and parts to the many logging camps and mills throughout the Northwest timber belt—gears, sprockets, locomotive shafts, brass fittings, and small machine pieces used extensively in logging and sawmill operations and all the allied industries. The Lamb people were pace-setters in heavy manufacture, from plans and specifications into the foundry, where the skills of chemist and melter combined to produce four grades of alloy, the hardest and most durable metals of the time.

Four hours after "500 horsepower" in electricity was turned into that great furnace, the metal batch was ready to pour. Rattlers, grinders, air chippers, and sandblast removed any foreign matter, slight imperfections were remedied by electric welding, and the castings were carefully checked for any serious weakness. Steel casting was undergoing constant changes and improvements, but far out on this remote fogbound Washington coast, the Lamb Company kept stride with the nation's best by using the latest methods of forging parts, and in its fully equipped machine shop of turret lathes, drills, a new horizontal boring machine, and an automatic gear and sprocket cutter, the company turned out heavy running equipment with the same pride and care of a Swiss jeweler.

During the Great War, as it was called, Uncle Sam sent in the Army to help cut spruce for Allied fighting planes. The troops were dispatched to the great spruce stands around Grays Harbor and Toledo on the mid-Oregon coast. Spruce manufacture demanded not only quantity but quality for the making of aircraft. The Allies wanted a huge air armada to knock out Germany, the lumbermen were informed at a cloak-and-dagger meeting at Grays Harbor, where precise specifications were laid out by government officials. To that time, spruce had been ignored in favor of the Douglas fir, but the lumbermen were told that at least 117,000,000 feet of usable spruce would be needed in the next year, meaning a total cut of about 1,000,000,000 feet of logs, since only about 167 board feet of every thousand met stringent government requirements for aircraft production.

That was only part of it, for in war, lumber is prime material, and if you ever wish to see a conservation movement blown sky high, start another war. The over-all cut ran to 790,000,000 board feet of Douglas fir for 208 ships of the Emergency Fleet (the government planned for an additional armada of 244), while the Navy consumed another 122,000,000 feet of lumber. The Emergency Fleet was built to specifications, as though stamped from a single mold, and the sawmills had to turn out the boards in exact lengths, widths, and thicknesses. But despite these hangups, the timber romped from the woods and through the mills so rapidly, to meet a frantic demand, that the boards were shipped green, and many vessels lost their calking on the first trip.

The spruce harvest also opened the way for other forms of manufacture, foremost being the E. K. Bishop Lumber Company, which specialized in fine piano sounding boards. Ned Bishop's father came from the Lake states seeking the coveted spruce, first near Montesano and then in 1920 shifting west to Aberdeen. The Bishops had long appreciated the value of spruce for special production. While others turned out plain fir boards, the Bishop spruce operation became a legend of The Harbor, garnering glowing admiration when it sailed through the Great Depression without losing a working day. Part of the key to postwar success was the Bishop plant's ties to the budding aircraft industry at Buffalo, New York, which grew into Consolidated Aircraft of San Diego; and certainly a symbolic peak was reached in 1927 when Charles A. Lindbergh rode into aviation history on his *Spirit of St. Louis,* built of Grays Harbor-cut Sitka spruce from the Bishop mill and the Posey Manufacturing plant at Hoquiam. When the nation's hero made his postflight tour of the nation, he dipped the wing of the famous plane over the place where its structural material had been made. It was a proud and exciting day along the banks of The Harbor, where even the skid road strumpets, who had little feeling for history, leaned out their windows to see the nation's hero fly by.

The color and excitement of The Harbor took various shapes, many of them off-trail slices of life. A. M. Simpson, the pioneer sawmill man who started things humming at Grays Harbor, never received any kudos or award as Citizen of the Year. But Captain Simpson was immortalized in a unique manner.

Among Simpson's employees before 1910 was a young bookkeeper who labored long, frustrating hours over the profits and losses of the Northwestern Lumber Company. Peter B. Kyne observed his two-fisted boss at close, critical range over a long period of time. Kyne liked working with words rather than with digits and decimals, and in Captain Simpson he saw a crusty yet lovable shipping tycoon who with hard-headed wisdom engaged in battles of wits not only with rival shipping magnates but also with his own shipmasters. Out of all this, Kyne created the redoubtable Cappy Ricks, among the most beloved and unforgettable fiction characters of the first half of this century. He also drew upon the rich flow of anecdotes and happenings of the lumber company and The Harbor.

Also exciting Kyne's interest was a tall, salty mate of the barkentine *Gardiner City* who sported a walrus-style mustache and a seasoned knowledge of winds, storms, foreign ports, and Erickson's establishment down in Portland. Ralph Peasley became a skipper, as in Kyne's books, but didn't marry the boss's daughter. Kyne's youthful hero, Matt Peasley, grew to international fame through the Cappy Ricks books and *Saturday Evening Post* stories, read avidly by millions during the glory years of the popular magazines.

When Cappy Ricks was born to print in the first of several hardbacks, the subtitle read, *The Subjugation of Matt Peasley,* which no doubt Captain Simpson found to his special liking. As with many budding authors who draw fiction from actual life, Kyne didn't know how other rugged individualists around The Harbor might receive the book, especially Ralph Peasley, so Peter protected himself with built-in insurance in a two-page dedication to his many lumber friends, and especially to "my good friend Captain Ralph E. Peasley who skippered the first five-master ever built, brought her on that first voyage through the worst typhoon that ever blew, and upon arriving off the Yang Tse Kiang River for the first time in his adventurous career, decided he could not trust a Chinese pilot and established a record by sailing her up himself."

The skipper swelled up like a balloon at being immortalized by the lumber bookkeeper. For some 60 years, Ralph Peasley rode the seas in a number of commands, the last being the schooner *Vigilant,* one of the many built right at Grays Harbor for the lumber trade. The *Vigilant* made several historic voyages to far ports with Peasley at the helm, and with Kyne's help, he became the most famous master of The Harbor. Whenever he hove into port along the Pacific Coast, waterfront stiffs would chant:

> My name is Matt Peasley,
> The mate of Cappy Ricks.
> I'm skipper of the *Vigilant*
> With all her five sticks.

However, Cappy Ricks didn't stand alone as the only fictional character from life in the Northwest lumber ports. He had competition in a "lady" whose yarn-spinning creator, Norman Riley Raine, drew from affairs on Puget Sound. Her name was Tugboat Annie.

Plank Island was a crossroads, as were the other lumber ports, for the tough lumberjacks and the sailors of the sea. In many ways they had things in common: the constant battle with the elements, the high risks, the lonely months of homeless isolation, the long hours and low pay, the lack of any decent working conditions, small hope for any kind of future. The waterfront was the key to nearly everything, for all movement depended on the waterways. The tallmasters, the steam schooners, the barks, and the barkentines tied up to the mill docks and processing plants, while the men sought the pleasures and sins of Hume Street. When the sailors were joined by the loggers on a holiday, the town shook as if from an earthquake.

Yet as with lumbering itself, Grays Harbor recognized that its very existence depended upon those vessels that passed across the dangerous bar, to 33 docks that could handle 55 ships at a time. The local citizenry organized the port to deepen the channels and improve navigation. Shipyards sprang up to increase the tonnage carrying the tremendous 30,000,000,000 feet of green gold being turned out by the mills. Many fine seagoing vessels were of native timber, some of them named "bald-headed schooners" because they had no topsails, like the *Andy Mahoney* from the Lindstrom yards, built in 1902. Until that time, the "teem kooner," as the Scandinavian boys called them, had a capacity of less than 400,000 board feet. Then Charles Nelson put together the *Nome City,* which had 1,000,000 feet capacity, and that broke things loose, the lumber ships growing bigger all the time. It spelled the demise for the windjammers, although the decline was slow, for they hung around a long while, their owners not realizing they were obsolete in the veritable rush of steam schooner construction and experiments with increasing capacities without expanding the size of the vessels. The Aberdeen Shipbuilding Company sent the 695-ton *Phyllis* down the ways, able to carry over 1,000,000 board feet; and

Frazier Matthews of Hoquiam built a number of noble vessels of over 1,000,000 feet capacity, among them the *Daisy Putnam,* the *Daisy Matthews,* the *Quinalt,* the *San Diego,* and the *Esther Johnson,* which was 208 feet long, 611 tons, and able to stoke away 1,275,000 feet in her hold and on deck.

That name Johnson haunted Grays Harbor, so there's no wonder it found its way onto the prow of at least one ship. At one time there were at least 14 Johnsons skippering vessels in and out of The Harbor; also so many Johnsons longshoring and working around the boats that to avoid confusion, a variety of nicknames sprang up—Rough Pile Johnson, Hungry Johnson, Swellhead Johnson, Drawbucket, Single Reef, Glassy Eye, Slabwood, Cordwood, Scantling Bill, Doughnut, and Scarface. There were too many Larsons, Andersons, and Hansens, also, so they acquired such handles as Breakwater Bill, Holy Joe, Salvation John, Fancy Ben, and Picnic Charlie, and the favorite of author Kyne, who used it in his stories, "All Hands and Feet."

The logging outfits and the sawmills branded their logs, as cattlemen did their cows. Log pirates gave a lot of trouble on the bay until a special harbor patrol was organized to drive hijackers from the log booms. Many of the pirates were presumed to be Wobblies, the "I won't work" crowd of the IWW, the "One Big Union," bent on wholesale disruption and sabotage, called anarchists and communists and reduced to pirating logs for ready cash.

Sleek sternwheelers plied the waterways, carrying mail, freight, and passengers between Montesano and Westport, and in summertime making special excursion runs for folk who had the time and wherewithal. The *Harbor Queen, Montesano, Cosmopolis, Harbor Belle, Wishkah Chief,* and *T. C. Reed,* burning slabwood in their fireboxes, ran up the Chehalis, loaded with gamblers in fancy vests and saloonkeepers with heavy gold watch chains; eastern businessmen in top hats and handsome attire; loggers, millwrights, ranchers, tradesmen, pimps, madams, strumpets, sea otter hunters, housewives, speculators, surveyors, tax collectors, thieves, card sharps, and even a preacher or two. Conversation in 1911–12 often turned to John Turnow, the strange Wild Man of the Olympics who was terrorizing the logging back country. Turnow was a social dropout who wished only to be left alone to live in the woods with the wild animals. But men wouldn't allow it, and so he killed those who hunted him down for capture, after issuing many warnings.

The fellow was a giant—six feet, four inches tall and 240 pounds of sinew and muscle—yet he could prowl the forests he loved without a sound and was so knowledgeable a woodsman that he could survive the freezing winters of the Olympic Mountains. Now there was a price on his head, $5,000, and bounty men were after him as a killer. Turnow set logging camps on edge, as he broke into cookshacks for supplies or his bearded face would appear at the window of an isolated cabin. Women feared to stay alone while their logger husbands were away. Ranchers moved their families to town. The Wild Man, as the stories built up, even frightened the loggers who didn't understand him. Once Emil Swanson was felling a big tree alone when Turnow swung from that very tree to another and disappeared in the woods. Swanson dropped his ax, rubbed disbelieving eyes, and tromped back to camp. He charged into the office, demanding his pay.

"By Yesus," he yelped, "aye vork 'ere no more!"

While milking or plowing, farmers packed their guns. Homes were securely locked, abandoned temporarily for the safety of the towns, and there weren't the normal spring fishing trips and picnics in the Satsop country. Turnow touched off one of Washington State's greatest manhunts, as from 1,200 to 2,000 citizens of three counties combed the wilderness. Finally on a tip, Turnow was located, trapped, and killed by Deputy Sheriff Giles Quimby beside a small lake deep in the Satsop wilderness, in what Quimby described later as the most eerie experience of his life. The body of the big, bearded Turnow was brought out to Montesano sitting upright in the car and placed on display, where 200 persons filed by to get a close-up view.

The tinkle of glasses beyond the swinging doors, the squeals of the girls, the rousing shouts from the saloons, the empty oaken kegs in the alleys, and the thick forests of tall masts in the harbor were signs of how things were going. Living was high and the bucks flowed freely as this last and greatest timber bonanza was wrenched from the western wilderness. Yet across town, beyond Heron, was another kind of society, of sober, churchgoing people, the solid citizens of Aberdeen who kept neat shops, opened schools, financed hospitals, and built ornate homes and fine mansions. The mill executives laid out streets and created building sites, attended the opera and musical entertainments and plays, traveled about in fancy rigs and then the early automobiles, and tried to have openly as little as possible to do with the waterfront district. However, a dollar being a dollar, Hume Street wasn't always off limits so they carted their wares down to the waterfront to sell to the girls and the saloon proprietors. The well-heeled girls often had fat "pungs," as the loggers called their pokes, and were susceptible to fancies, money being no object. Funny thing, they always paid their bills; strumpets like Lil White were good for a loan of a few thousand anytime at the local banks.

All got a piece of the Big Cut, and while it didn't have the sparkling appeal of Klondike gold, millions of bouncing dollars rolled up and down the streets, across the bars, and through the shops, markets, hotels, and all the rest, in and out of someone's pocket or purse. All of it came from those rich green veins of forest—those wonderful fat trees—of this amazing land. Now the logging outfits were pushing deeper and deeper up the Olympic Peninsula toward the high peaks, and there was a movement to put a stop to it. The nation was on a conservation kick, and politicians and do-gooders were sounding off about what was happening to this grand forest, this wonder spot of the world. They wanted to turn much of the 2,883-square-mile region, containing 61,000,000,000 board feet of merchantable timber, into a national preserve, maybe even a national park. Grays Harbor was horrified by the thought. The tri-cities were counting on that supply for years to come. If the peninsula were locked up, it meant the end of everything, since the sawmills were geared to handle the big stuff, not matchsticks, and that was almost gone. And then what would happen to Grays Harbor?

Chapter VI

Biggest Log Cabin in the World

Most original of all the exposition buildings is the Forestry structure. It is distinctly a western building and represents one of Oregon's greatest industries. . . . The building is the most original architectural creation ever offered to public view and is in many respects the most interesting structure on the exposition grounds. . . . It is easily recognized by its rustic structure.

OFFICIAL GUIDE TO THE
LEWIS AND CLARK EXPOSITION
June 1–October 15, 1905

When the centennial year neared of that now-remote time of Meriwether Lewis and William Clark, who floated with a small band of Americans down the broad Columbia River to the sea, people in the Oregon Country felt the need for some kind of celebration. They wished to call attention to the world that 100 years had passed, that their pioneer era—save for some fragments and romanticizing—was over, and that this far corner of the land was not only civilized and the Indians contained, but was growing rapidly and advancing soundly.

The idea of a World's Fair, first suggested in the mid-1890s by Daniel McAllen, a drygoods merchant, gradually took hold, especially among the business leaders of Portland, who were eager to promote their village, originally known as Stump Town. Their pressure on the state legislature moved that unpredictable body to approve an appropriation of $400,000, while local business interests pungled up $475,000, and then Uncle Sam came across with $1,775,000 to develop the world show. This was not without sound reasoning. The 1805–6 trek across the continent's wilderness had been not only a momentous achievement, but also helped significantly to bind the nation's destiny into a single package, coast to coast, and eventually to turn the tide against the encroachment of other powers, namely the hated British. Therefore, a centennial anniversary was worthy of an extravaganza at the far end of the Lewis and Clark trail, remote though it was and being sponsored by the smallest city ever to tackle such an event.

Two years went into the planning, excavation, and construction of this never-never land on a 385-acre tract in northwest Portland's Slabtown area, named because of the many sawmills and the mighty stacks of slabwood along the streets. And on June 1, 1905, in Washington, D.C., President Theodore Roosevelt, who had personally laid the exposition's cornerstone, pushed a button that swung open the gates to the sparkling grounds. As the crowds surged inside, it was as though they were entering another world, of graceful buildings, glittering fountains, beautiful gardens, and an exciting Midway, where Little Egypt danced and where you could buy for the first time a new confection of spun sugar called Fairy Floss, which later became known as cotton candy. You might also personally meet the Poet of the Sierras, Joaquin Miller, and the Northwest's bearded senior pioneer, Ezra Meeker, who would the following year retrace his steps by oxcart back over the Oregon Trail to mark it for future generations.

But among the beautiful buildings, one massive hulk stood out above all the rest, visible from most any point on the fairgrounds and certainly one of its prime attractions. It was the Forestry Building, dubbed the "biggest log cabin in the world." No one who saw it would challenge that fact. The Forestry Building was of such whopping size as to strike the imagination and boggle the mind. Paul Bunyan and Babe could move easily about its interior, while visitors wandering through the tremendous cathedral hall and side alcoves were struck at once by how small man is against the achievements of nature as displayed in the great trees shown here. It was a memory to be carried away to last a lifetime.

The great log house was erected with intensive pride, skill, and native inventiveness by the calk-booted, rough-talking lumberjacks and lumbermen of this thunderous logging country. More than 1,000,000 board feet of top-grade timber, the best that could be located in the Northwest woods on both sides of the Columbia, went into the gigantic building, 206 feet in length, 102 feet wide, and 72 feet to the highest pinnacle. A total of 52 matching virgin growth firs, six feet in diameter at the base and five feet at the top, marched the full length of this remarkable hall. The 54-foot main pillars—the largest weighed 35 tons—each contained 8,500 board feet of lumber, sufficient for a five-room house. There were eight miles of fir poles, a 184.5-foot flagpole, two miles of prime fir logs if laid butt-to-butt, 43,000 fir shakes, 30,000 fir bark hand-cut shingles, and in all, 1,200,000 feet of lumber. The entire building was estimated to weigh 32,640 tons.

Work was started more than a year in advance of the fair. Timber cruisers combed the Northwest forests for the finest specimens of trees to go into this Oregon Parthenon, as it was later called. A few were amazingly located within the city limits of Portland, not far from where the big cabin would be fitted together. Many grand firs were found on the Washington side of the Columbia, near the Oak Point camp of lumber king Simon Benson, who would make his mark sending tremendous seagoing cigar-shaped log rafts to his sawmill at San Diego. Other logs came from Pelton & Reed holdings near Goble in northwest Oregon timber country.

Since the building was to be a primitive work of art, not so much as a small bruise was allowed to mar the bark of these perfect specimens, which were to appear to the world as tall and fine as when they still lived in the Big Woods. To protect this facing, the loggers used tender, loving care by employing grabs instead of chokers to hook onto the butt end. The grab holes could later be seen by sharp-eyed examination at the bottoms of the upright pillars. At Pelton & Reed, none of the bucked logs ran less than 80 feet, and many were longer. Ira Withrow, who ran the yarding donkey, had the precarious task of bringing in the huge stuff, and this was before high lead logging. Withrow used a Tommy Moore block, designed in 1893 at Portland by Henry Hoeck. This heavy main line block, anchored to a tall stump, had a wide throat, and sheaves that allowed the butt rigging to pull through without fouling. Withrow got the sticks to the landing unscathed, but lifting them onto the railroad cars was another ticklish matter. They were too long to be handled by boom, so there was a sudden reversion to primitive methods. Two or three flatcars were run to the landing and the logs "parbuckled," using wire rope and steam power, to roll them down the skids and onto the wheels.

Despite all the time and care, many logs were discarded as "defective" because scarred bark couldn't meet the standards. As with Pelton & Reed, the magnificent sticks at Oak Point were six miles back in the brush, where they had to be felled, yarded, and hauled out, one to a car, by Shay locomotive, then rafted up the Columbia and Willamette rivers to Guilds Lake.

It took into June 1904 to get some 300 logs out of the woods. A $30,165 bidded contract was let, with $14,000 allowed for labor in the log placement. But raising the structure, symbol of the brawling industry that dominated the Pacific Northwest, was a labor of love for the lumberjacks. Many served voluntarily, something that couldn't happen today because the unions wouldn't allow it. When spring high water permitted, the logs were towed into the Guilds Lake area and winched by steam donkey from the lake to the hilltop site over a 1,550-foot chute. Then a 112-horsepower donkey and gin pole, or derrick, hoisted the big sticks into assigned position, the finest and most uniform placed on end in marching unison and the other heavy timbers resting horizontally one on the other to form the cabin walls. Foreman Ike Heisey handled the sticks with the industry's traditional know-how, carefully boring a hole near the top of each vertical column, then inserting a 2¼-inch steel shaft on which to attach the chain. Meanwhile, lumberjacks bitted, notched, hewed, and specially marked the various timbers for walls and superbracing, to be assembled exactly to specifications, and held tightly in position by their own size and tremendous weight.

A total of 3,000,000 visitors from around the world, in an age when distances were far-flung, came to see the exposition. At least a few were vaguely aware of Lewis and Clark, the prime impetuses behind the event. But they went away with a single common memory, for most if not all passed through the tremendous Gallery of Trees in the Temple of Firs, which was as big as the outdoors itself, and were overwhelmed by such things as a fully operating sawmill, exhibits of lumber and its many by-products, forest, fish, and wildlife displays, and panels of some of the first plywood ever manufactured.

There had been plywood, or "laminated wood," of sorts down across the centuries. The Chinese and Greeks had what was called "shaved wood." There were traces found in the tombs of the Pharaohs. Legend contends that the Trojan Horse may have been fashioned from what is now called plywood. But none equaled what appeared at the Lewis and Clark Centennial Exposition, made from the Northwest's own Douglas fir, the first pieces manufactured directly across the Willamette River from the fairgrounds at the Portland Manufacturing Company plant in St. Johns, which produced such inglorious items of wood as crates and baskets for the fruit and berry trade, coffee and spice drums, and excelsior.

Logging outfits and lumber companies joined together to develop exhibits for the great hall. As its part in this massive forest industry show, Portland Manufacturing was directed to provide an exhibit of "something new and different."

Gustav A. Carlson, who with his partner Peter Autzen, a Grays Harbor lumberjack, had begun the business three years before, had in his employ one N. J. Bailey, a skilled lathe operator who also had considerable know-how at manufacturing panels. There is no precise record, but the indication is that Bailey's knowledge of panelmaking might well have inspired the suggestion to attempt the manufacture of plywood as "something different" for the fair, the men having discovered how neatly the great fir logs peeled off into thin layers of veneer.

In March 1905, a special crew of six men, supervised by Bailey and Carlson, began the experimental manufacturing of plywood panels. Top-grade peeler logs were set aside for the St. Joe lathe and a steam kiln drier. They had no regular press, prepared glue, clipper, or sanding machine; and the experiment wasn't limited to the fir, but included other "Oregon wood"—alder, oak, ash, cottonwood, and red cedar.

The veneer was roughly cut without trim, then hung out to dry on a covered loading dock and placed in the kiln for two days of drying before gluing the sheets together. The gluing process was worse than working in a sewer. The animal glue used was so rank that workers applying it with hand brushes could hardly stomach the odor, dropping their brushes and rushing for the door to fill their lungs with fresh air. To make things worse, the glue needed to be kept constantly hot, for if allowed to dry, hours would be lost cleaning the pot for a new batch.

Work moved ahead at a slow, tedious pace. There were many failures. A wooden press was improvised with common house jacks and backing timber for support. Only a single set of paneling could be glued any one day, having to stand overnight or even longer. The fir sheets were found to be the best wood. Carlson and Bailey felt they were onto something, which indeed they were. Finally, Carlson happily announced successful production in *The Timberman,* trade bible of the lumber world, adding that there was already considerable interest in the fir plywood so that their efforts appeared not to be in vain.

During the fair's summer run, the personable Tom Autzen, sent to Portland by his father to assist with office and clean-up chores, showed and explained the panels in detail to half a million visitors, including many eastern lumbermen, who referred to the wood as Oregon Pine. Among the visitors were many door manufacturers, and by October, Superintendent Carlson was able to announce that "the company is building up a good business in glued veneer stock for factory use" and predicted with some modesty that the future looked "very promising," probably the understatement of the decade. A new wood-product industry was born at the exposition, soon to be manufacturing paneling not only for doors but for bureau and cupboard drawers and trunks.

Before long, plywood manufacturers were invading the Grays Harbor country, where the big peeler logs were in good supply. Peelers were hand-selected and carefully inspected by rolling them over a half dozen or more times before acceptance. One firm was soon shipping 50,000 doors per month to the United Kingdom and found a ready market providing flooring and running boards during the automobile craze of the twenties. Later, plywood was needed for aircraft, for construction of the San Francisco Golden Gate and Bay bridges, and in the manufacture of the PT boat fleet of World War II. The uses appeared endless, from children's play-yard equipment to wall panels and furniture, all stemming from those modest boards that made history at the Lewis and Clark Fair.

When the fair gate swung shut for the last time and the crowds faded away, the Lewis and Clark celebration suffered the same fate of all expositions of this kind. Buildings were torn down or purchased and carted away, many of them to other locales in Portland, where they long stood—and a few still stand—against the tides of the decades. But no one could cart off the Forestry Building, although a New Yorker thought he might. He wanted to buy the building for $200,000, dismantle it, and ship it East for a Coney Island attraction. Instead, it remained on its hillside site as a landmark to the fair and a reminder of the high pinnacle Portland had reached during that centennial summer.

The log cabin was presented to the city to be held in trust by the parks department, which nursed it through harsh weather, depressions, extensive use, and the effects of aging. Over the years millions of visitors to the City of Roses streamed through the gigantic structure, considered one of the region's most outstanding attractions. Hordes of schoolchildren came to view the lumber exhibits, updated from time to time by a citizens' committee headed by Thornton T. Munger, a veteran forest researcher, and guided by local lumber interests and the West Coast Lumbermen's Association, for the building continued to reflect the industry's "image." There was even a full-size forest lookout tower at one end of the hall that visitors could climb to get, peering through a fire finder down that breathtaking gallery of trees, some feeling of what it was like to be perched on a Northwest mountaintop, combing the ridges and canyons for signs of flames or smoke. Another attraction was a piece of 575-year-old Douglas fir, more than nine feet in diameter, which began growing on the Oregon coast more than a century before Columbus set sail for the New World. Visitors could personally count the rings as proof. And as the logging industry began to change dramatically, heavy old-time gear, including a set of high wheels and a retired logging locomotive, was placed on the grounds.

But the building itself remained the prime attraction. Later generations of Northwesterners and those from other sections of the country and the world who had never seen such timber, found beyond their grasp that this startling West Coast edifice could be built at all. People who were becoming conditioned to a world of plastic imitations asked attendants if the big timbers were "fakes." They found it difficult to accept such facts as the cost of $20,288.33 for construction, with the logs donated and the labor volunteered. In later decades of enlightenment, it couldn't be done, for such magnificent trees were no longer obtainable in quantity, and their economic value had increased many times. Labor and other construction costs would have made the project prohibitive, and chances were that the unions would have never allowed their members to work without full compensation, even if the lumberjacks wished to do so.

Several times, the big building needed repairs and renovation. In 1914, it was given a concrete foundation costing $6,000, donated by James J. Hill, the railroad builder. While most of the timbers remained sound after half a century, a few needed replacement from time to time to keep the building safe and to stay deterioration. After all, it was built to stand only 25 years, now was more than double that age, and was estimated good for at least another quarter century. Every so often, there were suggestions that it be leveled, that it was becoming a white elephant and a dangerous hazard, and that the lumbermen were weary of pouring in huge sums for its upkeep, since it no longer reflected their image in the modern world. But public sentiment was strongly opposed to its destruction, with outcries loud against any such suggestion. The course of events finally settled the matter.

During its fifty-ninth summer in 1964, fire as quick and total as any that ravaged the great forests destroyed the historic timber palace which, right to the end, was still drawing steady crowds, among them third-generation youngsters brought by their grandparents who vividly remembered the gay fair of their youth. The twilight holocaust exploded with a roar, flames shooting 150 feet into the sky, to be seen over a wide area of metropolitan Portland and the surrounding countryside. A total of 21 truck companies rushing to the scene were unable to save the Oregon Parthenon. The fire was too hot, the flames having a good head start and spreading too fast to control, for the old building's timbers were bone-dry. A safety sprinkler system did no good. The fire roared through the Gallery of Trees with a mighty thunder. The crackling and crashing of falling timbers sounded like the same kind of terror in the woods from which those giants had been extracted six decades before. All that firemen could do was contain the inferno by protecting surrounding buildings, including a large department store just down the hill.

One by one the tall columns, walls, and rafters that Ike Heisey had cursed into place fell amid showers of sparks, accompanied by the booming, thrashing, and tossing of splintering logs of an age gone by. All the building and its treasured contents, including the first plywood, were consumed. By morning, only the black skeleton remained, resembling a fire-ravaged wilderness of fallen giants, while streams of cars crept slowly past the scene as though attending the last rites of some beloved public figure. Tears came to the eyes of many who knew the building as a part of their own lives. There was indeed a deep sense of personal loss, not only among oldsters for whom the Forestry Building and the long-ago fair were something special, but among many others who saw the hall as symbolizing a glorious and robust time in the timber industry.

There were ugly rumors that the lumbermen themselves had destroyed the building, since they longed for something more in keeping with the times. There were tall timber stories of arson and of small boys seen running from the scene just before the building burst into flames. However, Jeff Morris, the reliable fire investigator for the city of Portland, squelched the stories. His conclusions were that the fire started from sparks of a refuse burner on the grounds where workers had been cleaning up. Someone must have tossed a flaming starter such as a cigarette or match into the container. Sparks flew into the air, lodging among the terribly dry logs.

"They may have smoldered there for some time before bursting into flame," Morris said. "The fire had a good start before anyone ever knew it."

In many ways, the sudden and complete conflagration was a far more appropriate end than the slow, sorry dismantling of the structure, which might have been its fate eventually. It belonged to an age in the Big Woods that by 1964 was fast fading into history, an age when timber and its people lived constantly with fire, more often than not of unknown origin. I couldn't help reflecting that had those giants of the Forestry Building never left the woods, they might well have suffered a similar end.

As a denouement, a few weeks later in that same tragic summer, the lumberjacks' Boswell, Stewart Holbrook, hung up his hobnails and passed from the scene.

For many years he'd lived and worked in the West Hills only a short distance from where the loggers' grand temple stood. Its finale must have been a great shock to him, for Holbrook had often brought visiting friends to see the magnificent hall, which symbolized his world. Certainly the double loss within such a short space of days underscored, as could nothing else, the end of an era.

There is an epilogue. Leaders of the timber industry realized that even though they may have not shed many tears over the log cabin, something was needed to tell the public the industry's story. It was impossible to duplicate the old building. After much discussion and planning, a new Western Forestry Center, depicting the modern techniques of logging, lumbering, wood products, and forestry was developed in Portland's West Hills at a cost of $2,500,000. It was dedicated in the summer of 1971, a beautiful structure, dramatically erected in an octagon-shaped central hall with eight finely fashioned laminated fir pillars reaching 70 feet to the sky-lit dome. As with its predecessor, the main building of a two-unit center (the other is a meeting hall) has huge four-inch double doors of hand-crafted vertical grain Douglas fir and hung in a framing of fir. Fir beams 18 inches deep support the foyer roof, while placed at right angles is 3-by-12-inch decking of ponderosa pine, to form tightly fitting tongue and groove edges. Large Douglas fir horizontal beams, 9 by 36 inches, support the balcony of the rotunda.

Thousands of 2-by-6-inch Douglas fir pieces were fitted together for the balcony floor. They were placed end-up, showing the sapwood, heartwood, and annual growth rings of the original trees, with their rich light-and-dark variations. The blocks are only 1.5 inches thick, but were pressure treated for durability that lumbermen say will even outwear the foyer and rotunda floor of quarried slate from Washington's No Name Lake. To the left and right of the stairs are A-frames housing moving models of a sawmill, pulpmill, and pulp and paper and modern logging. Out of view are offices finished in pecan and teak plywood sequence-matched walls, and a classroom of oak paneling and teak feature strips. To the west of the main-floor rotunda is a "memorial hall" of sugar pine from the mountains of southern Oregon and containing two contemporary wood carvings by Thomas Hardy of a forest scene and sawmilling. This particular Hall of the Tall Timber contains a black walnut cabinet holding the biographical stories of leaders in the timber industry and its development in the Northwest.

The many exhibits surrounding the rotunda and located on the balcony relate the timber story as it exists in the latter third of the twentieth century. An investment of over $800,000 has been sunk into the exhibit development. The hub of it all is a towering "talking tree," which answers questions submitted to a drop box in the foyer. You can't help but speculate that old-time lumberjacks, fresh off the skid road, would have a whale of a time with that one.

Special sections are devoted to telling about modern-day techniques—forest land use, soil and timber conservation, tree farming, timber harvesting, reforestation and forest management, fire protection, wildlife conservation, and the great array of uses for wood and forest products. As with much else of this supermodern age, the message comes across in film, recorded voices, flashing lights, and plastics, all activated by pushbutton. One of the most impressive, to me at least, is the sound and fury of a mounting forest fire, in all its sheer drama and destruction, leaving in its wake a wasteland. The campaign to keep the public alert to fire dangers goes on and on, ad infinitum.

Most everything to get the lumbering-forestry story across to modern Americans who obtain their education from television is handled in the fashion of these times: "Teach me *but entertain me*." This isn't primarily a historical museum of logging and lumbering. Those may be found near Klamath Falls and Eureka. Only a small segment of the fixed exhibits touches on timber's robust years and the life-style of the sawdust savages, handled in a light manner through moving cartoon cutouts. The fallers and buckers of yesteryear would likely be disappointed, or merely amused by it. Yet in prowling about the center, they'd soon discover a growing number of familiar artifacts of their own heyday, the tools that they'd used, and some of the bigger rigs, including logging high wheels and the wonderful old One Spot Shay locomotive of 1909, which I saw in action in the 1940s at Stimson's sawmill and which later stood outside the old Forestry Building. By some miracle, it survived the fire, to be refurbished and moved in 1972 to the new forestry center. And on the balcony level is a display of pictures and text telling the story of the Forestry Building and its amazing construction.

The center, which attracted 200,000 visitors in the first 18 months and is the only one of its kind, hasn't relied wholly on simulation. Located as it is in a parklike setting of timber, near the famed Hoyt Arboretum of 600 varieties of trees, some of the 2,000 volunteers who help run the place are able to conduct nature walks during the summer months. There are also organized bus tours to tree farms, the new Tillamook forest, and the important forest laboratory at Corvallis. All told, it is an active endeavor, keyed to the now, yet a sharp break with the past, as are many things today, and an effort to place those wilder, freer times in some kind of perspective without outrightly rejecting them.

"We are trying to represent this scientific age," explains Ernest L. Kolbe, who came out of retirement as a career forester in the timbered West to manage the project and see to its development.

It is one fine building, for sure, with its broad representation of Northwest woods, not only Douglas fir but western red cedar, redwood from the northern California coast, white fir from central Idaho, sugar pine of southern Oregon, West Coast hemlock, and other varieties. More than 700,000 board feet of lumber, shakes, and plywood went into the place, selected from 36,000 pieces of lumber. The columns are sleek, the walls and floors highly polished, and the grains of these woods possess a special beauty of nature's own handiwork, unable to be duplicated in quality by machine, chemistry magic, or computer.

Still, whenever I go there, my mind reflects on the rustic grandeur of the old Forestry Building. The chances are that no corks will ever echo through these highly polished halls and carpeted stairways, and I wonder at the reaction of some rugged backwoodsman (there are still many around) on first entering the foyer. Shucks, a fellow wouldn't dare let fly with any four-letter words around here, or spit on the floor. Too much like a church. And I think he would long to sink an ax into those sleek columns, just to learn if they were real. Then he might turn in his time and hit on out, for it would be too overwhelming, reminding him that he was the product of an age that had all but disappeared.

As Director Kolbe explained, the new center is tied to the new age of lumbering and forestry, as the Oregon Parthenon belonged to another kind of world. It's too bad that it couldn't have survived, for both have their place, and the world's biggest log house can never happen again.

Chapter VII

Baths and Clean Sheets– for Loggers?

We want the workers of the world to organize
Into a great big union grand
And when we all united stand
The world for workers we'll demand. . . .

We want the timer and the skinner and the chambermaid
We want the man that spikes on soles
We want the man that's climbing poles,
And the truckers and the muckers and the hired hands,
And all the factory girls and clerks
We want everyone that works
In one union grand.

*IWW Songs to Fan the Flames of
Discontent*
Songs of the Worker
Joe Hill Memorial Edition
Published by IWW, Chicago, 1917

The woods were changing, and in the end would never be the same again. The thickly forested back country of the Pacific Northwest was emerging, despite hidebound resistance and a love of individual freedom, from the age of the pioneers, when anything went in the deep woods and it was every man for himself in the matter of survival.

Mechanization was received in the woods with much enthusiasm, for the steam donkey and the geared locomotive helped make it easier to get the big sticks to the mills faster and with less frustration. Within a few short years, the internal-combustion engine would be invading the forests with its wheeze and clatter, and when the woodsmen found that they could bring out the logs with gasoline-driven rig—trucks, tractors, and tractors with tracks—old Henry Ford himself made a trek to the Pacific Northwest to learn just what they were doing with his engines.

But the woods still demanded a lot of musclepower, you still had to be tough and rugged to fell and buck trees, high climb the tall spars, set chokers, and handle rigging as thick as a cobra. The felling and bucking arena would be the last holdout against mechanized power, so that lanky lumberjacks like a certain Swede up in the Grays Harbor country never need fear for their jobs. That faller was a working idiot, a legend of the area, who had no chin, so that they called him Andy Gump, a cartoon character of the times. He could stand on a springboard all day, working that misery harp biting through a huge fir or spruce with the steady rhythm of a water pump, first using one great arm and then the other, and never missing a stroke till quitting time. At noon he'd grab a fat sandwich in one big paw and keep right on cutting while he gulped it down, a human dynamo who wore out many a partner with his remarkable endurance.

Men like Andy Gump would eventually fade from the scene, although it took more than half a century to accomplish the full revolution. But when it was achieved, it was so complete that second-generation lumbermen and foresters voiced their utter amazement. The signs of future change in both methods and attitudes were evident at the Lewis and Clark Fair and were even stronger four years later, when competitive Seattle me-tooed its sister city to the south with its own world's fair, the Alaska-Yukon-Pacific Exposition, to celebrate the brawling rush to the frozen North by the Klondikers of '98. This fair had a forestry building, too, somewhat similar to the big log house built in Portland but never to stand as long as a Seattle landmark. It was a repeat performance of the showplace of 1905—no one could come up with a better idea—and thousands explored the big building.

One particular group, coming together for the first time, had their photograph taken before the logging center, although only half of the 100 plus in attendance showed up for the picture. This first annual "educational" gathering of the Pacific Logging Congress brought together tough logging operators in an atmosphere well removed from the skid road surroundings generally associated with the woodsmen. Sessions were held in the nearby House of Hoo-Hoo, built by fun-loving loggers for the exposition as some insurance against too much dullness connected with this Alaska-Yukon-Pacific celebration, although with all the Alaska sourdoughs running around, how they arrived at this conclusion is difficult to understand. Hoo-Hoo was the lumbermen's fraternity devoted to fun and games. But just what went on during those historic three days of the Logging Congress, which was billed as a "major event" of the exposition, was lost to history, though not through any lack of foresight on behalf of the founders of the Congress. They hired an itinerant tramp newspaperman to record the formal sessions in detail. Unfortunately, the fellow didn't have the necessary follow-through, except when it came to nipping steadily on John Barleycorn. At the end of the three days, he picked up his pay and turned in his report, which consisted of two classic lines, listing the name and address (both correct in the pride of his profession) of George S. Long, Tacoma, Washington.

Nevertheless, the congress was a huge success; the loggers found value in getting together. A second was staged the following year in Portland; it was then to become an annual affair, a significant step by an industry that was one of single-footed independent action, going off in all untamed directions. The idea was largely the brainchild of George M. Cornwall, the brilliant founder and editor of *The Timberman,* which would become one of the industry's leading trade publications. The suggestion grew from a conversation about baths for the crews between Cornwall and Ed G. English, a prominent lumberman from Mount Vernon, Washington, who pioneered the use of wire rather than rope in the woods. Baths for loggers? Such a thought was enough to make old-time lumberjacks go out and bite a tree.

There are several versions on how this all came up, and it doesn't matter save that it was a very hot summer day. One has the two lumbermen striding along Seattle's First Avenue, another finds them seated in the Frye or Diller Hotel, sipping cool drinks and deploring the humid day. Cornwall, who was miserable, was thinking of a bath.

"Ed," he mused, "wouldn't it be wonderful if we could provide baths for the loggers? They get so hot and dirty in the woods."

English may have choked on his drink, for not many in the business ever expressed much concern for the timber beasts. But Uncle Ed agreed that it was a good idea. The course of their talk took the whole notion much farther to the thought of getting the operators together occasionally "where we can bring this and other matters up for discussion." Cornwall began organizing the Logging Congress through his magazine, and among the significant topics of that first gathering were the problems of hygiene, camp pollution, and bettering conditions for the workers, including bathhouses. Operators were coming to realize that disease, particularly typhoid, was taking a heavy toll on their employees. There were two other unusual items on the agenda: forest-fire prevention and conservation or reforestation. That such matters would even be brought up for discussion at a gathering of freewheeling loggers would have old-timers digging wax from their ears. But visionaries like Cornwall were sounding warnings that are still being heard more than half a century later.

"The paper by Dr. [W. C.] Belt on camp hygiene introduces us to the broad field of sanitary engineering," observed timberman Frank H. Lamb following that particular discussion. "And to many of our old-time loggers, it may seem a little fastidious."

Cornwall editorialized on the matter of conservation: "It is a characteristic of the American people to lock the stable door after the horse has been stolen. Prodigality of her natural resources is a conspicuous example of this. Her natural gas, minerals, and forests have always been used with reckless disregard for the future and seeming oblivion to the fact that they are not inexhaustible."

The Pacific Logging Congress was a sign of the times, born in a period of great sweeping consciousness and feelings of guilt by the American people toward the nation's natural resources and their preservation, as prompted by Theodore Roosevelt, who held a great love and concern for the vast outdoors, particularly the West. In the budding years of the new century, attitudes and ideas were taking a new view. Loggers and sawmill workers of a young generation were beginning to stir for better and safer working conditions, for the accident rate in the woods and mills was appalling, disease in the camps frightful, and the general using up of good men long before their time seemed inexcusable. Despite the slight mutterings being made at meetings such as the Pacific Logging Congress, it would take force and violence to get across the message to many operators. Many an old-time boss lumberman, who'd come up through the sorry ranks of the tall timber at high risk and long hours, sincerely believed that loggers liked things rough-and-tumble, that they didn't want clean sheets on their beds, hot baths, and bunkhouses free from lice, bedbugs, and rodents. The lumberjacks were illiterate, footloose rabble, ignorant and "hardly human."

Two years after the Logging Congress met, as the festival-loving Northwest observed another centennial over the arrival of the men of John Jacob Astor to establish the first American outpost in the Far West at Astoria, several hundred lumbermen from Oregon, Washington, and British Columbia gathered at Raymond on the Washington coast, south of Grays Harbor. Many of them were wealthy and, like the loggers, god-awful independent. A reporter scratching on his notepad estimated that 11 of them represented the combined wealth of $100,-000,000.

They were called together by Everett G. Griggs, a powerful Tacoma lumberman who 10 years earlier put together a package for the Puget Sound sawmills —the Pacific Coast Lumber Manufacturers Association—so that as Cornwall had suggested for the loggers, the lumbermen could pull together on common problems and stand as one against the Wobblies and other union movements. Now Griggs hoped to do the same thing on a much broader scale, especially in dealing with railroad shipping costs, establishment of sound grading rules, local, state, and federal taxation, and the promoting and marketing of the region's lumber products around the world. But trying to bring together to a common mind such a gang of rowdy, outspoken, and freewheeling mill operators took no little skill and a powerful set of lungs. Nevertheless, from the Raymond gathering was born the West Coast Lumbermen's Association, which lasted half a century and had far-reaching impact on American thought, culture, institutions, and living patterns. WCLA became influential on many fronts, from making the public readily aware of its responsibility in fire carelessness to the full-fledged promotion of the tree farm movement. Yet in the beginning it was touch-and-go, for the lumbermen were their own worst adversaries from that maverick pride of independence.

"Every time a price is given," observed W. B. Mack, an Aberdeen operator, "some other fellow cuts it. Everybody seems to be fighting everybody else."

Troubles with the men were mounting and soon would overshadow everything else. The attitude of the industry was one of little responsibility; if you didn't like an operation, you could draw your time and move on. Cyrus Walker of Pope & Talbot at Port Ludlow and Port Gamble demonstrated the employer point of view when a man was crushed by a sliding load of lumber. The hat was passed for the widow, but Walker refused.

"I would like to help that poor woman and her children," he declared, "but I can give nothing. If I did, it would set a dangerous precedent and give the impression that an employer has responsibilities to his employees."

The workers didn't see it that way; while employers talked of bettering camp conditions and even of workmen's compensation, the bindle stiffs felt the bosses needed more than a gentle nudging. The great mistake made by the operators was a complete lack of understanding. They were certain the men didn't really want comforts and safer conditions, that they enjoyed the rugged existence and all the risks that went with the job. The installation of hot baths and sheets on their bunks would be insults to their manhood. The operators read the loggers wrong. Despite their crusty outer reputations, the jacks were human, with all of mankind's cravings and yearnings for a better life. Significantly, once when the Young Men's Christian Association set up an experimental program at a logging show during the Fourth of July holiday, all but a half-dozen loggers stayed in camp rather than hike down the road to a small settlement where saloonkeepers and strumpets were looking for a killing during the holiday.

The men couldn't rely on the generosity of the logging operators and lumber companies to better conditions. Any progress would be accompanied by the violence of destruction, beatings, killings, and cries of "communism" so characteristic of any effort to alter the status quo on the American scene. The workers turned to unionism, realizing that only through a combined effort would working conditions be improved. The shingle weavers took the lead, even before the turn of the century, because of the nature of their dangerous craft, centralized as it was, plus the fact that the shingle weavers were a rather homogeneous group. But bad timing of a strike in a depression destroyed their first union. In January 1903, the locals regrouped to organize the International Shingle Weavers Union of America. That same year the sawmill workers were beginning to organize also; two years later they merged into the International Brotherhood of Woodsmen and Sawmill Workers, chartered by the American Federation of Labor.

The strikes came on, with bitterness on both sides. While visionaries talked and read papers at their annual gatherings, favoring improved working conditions, many boss loggers and lumbermen set their calks solidly where they stood, firmly opposed to any changes or to the raising of wages and the shortening of working hours. They'd come up the hard way and balked at being told how to run their affairs. Meanwhile, the shingle producers continued trying to regulate prices, control production, cut wages, and buck organized labor. Most of the strikes were local, but in 1906 at Ballard, Washington, the ball game changed when the manufacturers met, raised $50,000, and outlined a campaign against the shingle weavers' union. The fund was for the protection of any manufacturer involved in a labor dispute and to provide for workers, essentially strike breakers, at the Ballard mill.

The angered shingle weavers retaliated with a strike against 365 mills affiliated with the Shingle Mills Bureau, but the manufacturers outmaneuvered them by opening a plant in a single locality, thereby drawing paycheck-hungry workers from other mills. It forced the union to capitulate, since its own membership wouldn't hold solidly together. But six months later, the union was again demanding higher wages, and this time won them without a strike, since the market had vastly improved and employers didn't want to shut down the mills. It was characteristic; throughout the long history of logging and lumbering, the industry has been extremely sensitive to existing conditions and likely to react in sweeping fluctuations to momentary changes of attitude, demands, and the ready market.

Yet it was the invasion of the Industrial Workers of the World into the Pacific Northwest that created the greatest turmoil and tragedy, and led from one crisis to another. One Big Union . . . the Red Dawn . . . the IWW . . . the "I won't works" . . . the sign of the Black Cat . . . the Wooden Shoe . . . American syndicalism . . . these were the Wobblies, organized "back East" in

Chicago in 1905 and acquiring their nickname from an Oriental restaurant-keeper in British Columbia who became tongue-tied over the letters IWW, blurting instead "I Wobbly Wobbly." The Wobs lost little time making inroads into the western mines, and then they invaded the Big Woods. They were feared as dangerous radicals and dyed-in-the-wool communists in the eyes of industrial, business, and social leaders who spread the charges against the IWW. There was bound to be trouble wherever they emerged, often under the direct and calculating eye of their leader, Big Bill Haywood, who was tough and fearless. He had manpower standing behind him and thousands of workers ready to answer his bidding when he called for support of strikes anywhere in the country, so that even loggers from the Columbia River show bobbed up in Massachusetts, where Big Bill needed them in a textile strike.

The Wobblies appealed to the long-suffering Northwest loggers because of their bold, aggressive methods at a time when things were primitive and intolerable in the woods. In short, the IWWs were activists.

"The working class and the employing class have nothing in common," IWW leaders stressed time and again. It said so, too, in the little red song books and other IWW publications:

"Make it too expensive for the boss to take the lives and liberty of the workers. Stop the endless court trials by using the wooden shoe on the job. . . ."

"Don't forget that a short workday, and big pay, always go together. . . ."

"Eight hours a day would put thousands to work. . . ."

"Our fight is your fight. So let's fight together. . . ."

"Every worker should have an ambition to live to be a healthy old man or woman and hear the whistle blow for the bosses to go to work. . . ."

"Union scabs—My dear brother, I am sorry to be under contract to hang you, but I know it will please you to hear that the scaffold is built by union carpenters, the rope bears the label, and here is my card. . . ."

"An injury to one is an injury to all. . . ."

"One union, one emblem, one card . . ."

Such statements, spelled out, struck home to the lumberjacks, looking at their meager surroundings and remembering the latest logging accident where a buddy was crushed to death while the lumber boss sat safely behind a desk in a warm office. They remembered, too, their ignored pleas for improved conditions and echoes of the words of Cyrus Walker about setting a dangerous precedent. The IWW advocated not only organizing the working people into One Big Union, but the overthrow of capitalism and the establishment of a system of industrial democracy.

"War is hell," cried the leaders. "Let the capitalists go to war to protect their own property."

Under the circumstances, the lumberjacks had nothing to lose.

The Wobblies created turmoil and terror, then moved in whenever a crisis arose. The brand of Marxism was upon them, and there were avowed communists within their ranks—Haywood himself later fled to Russia, where he died, and he is buried in the Kremlin. But the bulk of the lumberjack membership went along with them because that's where the action was, and it was a means of retaliating against the ills and cruel methods of the freewheeling industry. The Wobblies fomented strikes and riots in the lumber centers, preached hate and revolution, challenged the establishment with free speech movements, set fire to forests and sawmills, dynamited businesses, created riots in log drives in Idaho and Montana, and spiked logs at the mills—a terrible thing when the high-running saw struck the iron, with metal flying like shrapnel all over the plant. Yet they advocated progressive changes that the logger wanted and deemed necessary for his continued survival, among them safety precautions and regulating laws, an eight-hour day, and a $3.00 per day minimum wage.

The "I won't works" made their first great mark in the Northwest by grabbing the initiative during a 1907 Portland sawmill strike of independent workers. The American Federation of Labor denounced both the strike and the Wobblies. This lack of support by the AF of L gradually got the sawmills operating again, as workers drifted back to their jobs. And while the Wobblies lost the game, they gained public notoriety and demonstrated to many bindle stiffs that this union had guts. Moreover, wages were increased, through no doing of the Wobblies, who nevertheless took a share of the credit.

The Wobbly movement grew steadily; more and more timber workers were carrying the red cards and paying the $2.00 to $5.00 dues. It wasn't long before the movement hit the Big Bonanza country at Grays Harbor, where the IWWs were up against some of the toughest people in lumbering. There were six strikes at The Harbor in 1912 over the simple demand for higher wages, beginning at Hoquiam and spreading quickly to Aberdeen and Raymond, even though the Aberdeen Trades Council, which was affiliated with the AF of L, declined an endorsement of a $2.50 minimum and eight-hour day. Tempers flared, there were beatings with axhandles and pickhandles, and threatened lynchings of the strikers. Vigilantes decided that the best way to resolve the matter was to run all the Wobs out of town. Raiding IWW headquarters, they took custody of the leaders and rounded up every suspected Wobbly in Aberdeen. At Raymond, a force of 460 "deputies" corralled 150 Wobblies for boxcar shipment out of the area, while at Hoquiam a mob dragged about the same number from their homes and boarding houses, stuffed them into boxcars, and sealed the doors to send them on their way. But the railroad company wouldn't move the cars, and the mayor opposed the distasteful scheme; at the last moment the deportation was halted by the sheriff, and the alleged troublemakers were released. It was a sorry matter, for many had been badly beaten, and one was crippled for life. Nevertheless, a deportation plan of sorts was initiated by vigilantes to drive the Wobs from the Grays Harbor area at the time when the IWW was trying to spread the strike throughout the state. Increased wages, a preferential union shop, and reinstatement of all the men were the key issues. The sawmill operators adopted a proposal suggested by the Aberdeen Citizens Council for a $.50 raise to a $2.25 minimum wage and a 10-hour work contract, preference was expressed for the AF of L, and all Wobbly members were excluded. The loggers received a matching increase in pay and also milestone achievements: springs and mattresses in camp bunkhouses, and a clean-up of unsanitary conditions. Since crews were being secured anyway, and some of the strikes were drifting back, the walkout was lost.

In the next few years the Wobbly movement, and indeed most unionism, entered a decline, but World War I gave the Wobs impetus for new causes of disruption and agitation against the "capitalists." However, the picture had changed; the woods and mills were reorganized into other unions. The shingle weavers retrenched into their original rather exclusive "club," for their leaders were discouraged over attempts to organize the hodgepodge lumber industry on a wide-scale basis.

Hate against the Wobblies grew in intensity, so that there are men and women to this day in the Northwest who see fire at the very mention of the name. The Wobs were beaten, thrown into mill ponds, pistol-whipped, lashed with knotted rope ends, coated with hot tar and run out of town, knifed, and even hanged. It was war within the war, for the Wobs as a secret society infiltrated the ranks of the AF of L, particularly the shingle weavers, where they could instigate "quickie strikes," keeping things in an uproar. The fury in a period of intense wartime emotions and appeals for patriotism was bound to end in tragedy, and it did, when Wobblies decided to "open up" Everett, Washington. The "free speech" issue was interjected to provoke a crisis, as it had earlier at Spokane, Wenatchee, Walla Walla, Aberdeen, and Seattle; and as it was used a half century later by another generation in the late 1960s on university campuses. Certainly, it is a remarkable phenomenon where the matter of free speech, which Americans rate as one of their basic freedoms, can be used as a technique to rally sympathizers to a radical cause.

Forty-one Wobbly supporters left Seattle by boat on October 30, 1916, bound for Everett "to take over the town" during a shingle weavers strike. But vigilantes led by the local sheriff met them at the dock, gave them a sound beating, herded them to the edge of town, and then forced them to run a gauntlet, to be beaten bloody with clubs and sticks.

There were no serious or lasting casualties, but neither was that the end of it. A month later, promoting the same free speech cause, card-carrying Wobblies and their supporters returned to Everett numbering one less than 300 strong. A total of 260 were aboard the *Verona*, the regular passenger steamer, while the rest came on the *Calista*, a chartered ship. The word went ahead of them; 200 armed vigilantes confronted the Wobblies at the pier. As the *Verona* brushed the dock, a deckhand tossed a mooring line around a bollard.

Sheriff Donald McRae was a former shingle weaver but was now against the union and the hated Wobblies. McRae demanded the names of the leaders. The invaders shouted refusal. Then the shooting started. Which side fired the first shot wasn't known and it really didn't matter, for men dropped on the ship and on the pier. The crowd rushed to the other side of the *Verona* to escape the flying lead. That threw the vessel off balance and straining at its mooring line. The railing snapped and some men fell into the water, drowning because they couldn't swim and no one was able to help in the surrounding turmoil. Then someone cut the line and the ship backed off, out of gun range. But seven were dead (two ashore and five aboard ship) and 50 others were wounded (31 Wobs and 19 vigilantes). It couldn't be determined how many drowned. However, the Wobblies later contended that the men aboard ship didn't have guns and the wounded vigilantes were shot in crossfire by their own men.

When the *Verona* reached Seattle, 74 suspected Wobblies were held on murder charges. Emotions ran high, but although threatened with recall, Mayor Hiram Gill joined with the Seattle Labor Council and the State Federation of Labor in denouncing the Everett "authorities." The first man tried, Thomas H. Tracy, was acquitted, and eventually all were released for lack of evidence. But Big Bill Haywood's flamboyant Wobbly press wouldn't allow the issue to die, calling it the Everett Massacre; and I can only guess what the national impact might have been had the tragedy been covered in detail and channeled to millions of American homes by the medium of television, as were the riots and shootings of the 1960s. But this was 1916 and also the very far corner of the land, well removed from the center of the nation and so misunderstood that easterners believed the Indians might still break out at any moment, there were few paved roads, the people were generally illiterate and ignorant, and the only thing they knew how to do was cut trees.

Now the cannon of the War to End All Wars was growing louder, and the United States was ready to throw its strength into the conflict. While there was no love for the IWW among the AF of L unions, all the timber unions had common objectives. The time seemed right, unpatriotic as it might appear, to press for their demands, so the Shingle Weavers and Timber Workers unions, 2,500 strong, got together to foster a regional strike. The Wobs were organizing, too, and with this more radical element, it mattered little that the nation was at war. The demands were specific: higher wages, better camp conditions, union recognition, and the eight-hour day—this latter becoming the critical issue with the lumbermen. If employers didn't settle, the walkout was set for mid-July. But the operators refused to budge, organizing instead the Lumbermen's Protective Association, which raised a $500,000 strike fund and a $500 per day fine against any member operating less than 10-hour shifts. Linking their stand with that of national defense, the lumbermen asserted that "the establishment of an eight-hour day at this time is impossible, and employers hereby pledge themselves unequivocally to maintain a 10-hour day for the purpose of maintaining the maximum production in the lumber industry."

The men hit the bricks. Eighty-five per cent of the region's logging camps and sawmills were closed down by the greatest mass tie-up of the industry known to that time. The walkout was especially bold, and branded irresponsible, since it came in the face of heavy demands by the military, especially spruce for airplanes. Operations were down for six weeks as some 20,000 workers supported the strike. The central issue was the eight-hour day; at Grays Harbor even ships' carpenters refused to handle lumber from 10-hour mills. The governor of Washington issued a proclamation urging employers to accept the shorter day, but the lumbermen held their ground. A few independent operators gave in, but their capitulation was short-lived as the association applied pressure to its members not to supply logs to those mills.

By September, the demands were still unmet, but the unions' hold on the men began to slip away. Workers were drifting back to their jobs. However, the Shingle Weavers and Timber Workers refused to call off the strike until the shorter day was universally accepted. Even IWW leaders realized they could no longer hold the men back from their jobs; the need of paychecks and the pressures of public opinion of a nation in wartime were convincing arguments against continuing the strike. As an alternative, the leaders urged slowdowns and sabotage on the job. The grinning black cat, seated on a wooden shoe, might crop up anywhere as the symbol of sabotage. The Wobs were outlaws—it was unlawful to carry a card—but that didn't halt mysterious fires, railroad spikes in logs bound for the headsaw, derailed logging trains, and excessive waste of the raw product. Woods crews might work a few days, until the operator had stocked full provisions, then all quit; or halt work at the end of eight hours, forcing the boss to fire the bunch and hire another crew, only to have the same routine over again. Men didn't show up, shortening the crews and causing a reduction of the output. Anything for harassment and disruption. The operators were at a loss on how to deal with the situation, for the Wobblies were so elusive that you couldn't find them or their leaders, even if you had suspicions.

"It is almost impossible to deal with them," declared Senator William A. Borah of Idaho. "You cannot destroy the organization. That is an intangible proposition. It is something you cannot get at. You cannot reach it. You do not know where it is; it is not in writing. It is not in anything else. It is a simple understanding between men and they act upon it without any evidence of existence whatever."

Borah had it right; there was a fierce loyalty among the Wobs following the leadership of men who went by strange names, such as Horse Blanket Blackie, Shortline Red, and Boxcar Shorty. Decades later, an old-time Wobbly, with tears in his eyes, remembered the movement as the "organization for the workingman" with an intense loyal spirit among the members who "were clean, yet broke, but always happy and always carefree." The IWW was viewed as an unknown enemy within the country, with a tinge of communism and socialism, flying a red flag, and having red membership cards and a *Little Red Songbook*. Ironically, the hated Wobs contributed to the lighter side of the American scene with songs like "Big Rock Candy Mountain," and Haywire Mac wrote "Hallelujah, I'm a Bum," which was recorded by Victor Records, becoming a national hit in 1927.

The personal prejudice against the Wobs and what they stood for was so strong within the lumbermen's organizations that the operators wouldn't consider changes just from principle. They were clinging to the past over the eight-hour day in an effort to hold onto the kind of independence enjoyed for decades since the first settlers. Their aim was to break not only the IWW but also all the unions.

"This is no time to talk hours," declared President Arthur L. Paine of the West Coast Lumbermen's Association, as Uncle Sam shipped his doughboys overseas. "I almost regret that I have lived to see the day that my country would permit men to go unhanged who would go about spreading sedition at a time like this."

The Mediation Commission spelled out the trouble for President Wilson:

"This uncompromising attitude on the part of the employers has reaped for them an organization of destructive rather than constructive radicalism. The hold of the IWW is riveted rather than weakened by unimaginative opposition on the part of employers to correct the real grievances. . . . The greatest difficulty in the industry is the tenacity of the old habits of individualism."

The need for airplane spruce became critical, and the great plump Sitkas of the Pacific Northwest a prime source, described in noble terms by Major R. Perfetti, head of the Italian Military Commission for Aeronautics in the United States, as "a wood . . . which is consecrated by the Creator to insure liberty of the world and is the harbinger of peace and good will among mankind." Perfetti was thinking of airplanes in armada quantities to win over the Kaiser. The problem was to get that spruce, which grew to stupendous sizes along the Northwest coast, to the plane manufacturers in proper form in the shortest possible time. The Wobblies' storm and concurrent labor problems were something the government couldn't tolerate, for the war could be won or lost on the spruce supply, which could knock the Germans from the air over Europe.

Colonel Bryce P. Disque came West to investigate the production problems of the Northwest sawmills and to learn what all the strife was about. Disque concluded that a special force of manpower would be required to meet the government's need for 100,000,000 to 170,000,000 board feet of acceptable spruce. This meant that some 1,000,000,000 board feet must be cut, for only about 167 board feet of every thousand would meet specifications for airplane manufacture. At the time shipments were only half the monthly need. Disque assumed control in an attempt to organize the disjointed, bumbling industry of angry, frustrated lumbermen, then brought in the troops to get out the trees, and introduced a new government-backed organization called the Loyal Legion of Loggers and Lumbermen, known as the "4-L," to offset the quarrelsome unions and the Wobbly activities and to place the industry squarely behind the war effort.

By the end of 1917, the 4-L had 10,000 members in 300 locals. By midspring 1918, there were seven times that number in a balanced representation between employers and employees. The 4-L asked workers and lumbermen to scuttle their differences to win the war, all taking the membership pledge "to faithfully do my duty by directing my best efforts in every possible way to the production of logs or lumber for the construction of army airplanes and ships to be used against our common enemies. That I will stamp out any sedition or acts of hostility against the United States government which may come within my knowledge, and I will do every act and thing which will in general aid in carrying this war to a successful conclusion."

The idea of the 4-L may not have been merely Disque's, but may have originated with President Wilson and Samuel Gompers, or even among some of the calmer lumbermen who were anxious to win the war. The workers were suspicious at first; they'd been duped by logging and lumber bosses in the past and heard many promises, yet their working conditions, especially regarding safety, were as bad as any in the nation. Nevertheless, the 4-L enabled the two factions to come together to debate common problems—"communication," they call it today—and by the following year, the membership rolls hit 100,000 members. While Disque considered the 4-L a wartime activity, destined to go under when peace came, it lasted for years beyond the Armistice as an association of employees and employers. At times it was highly controversial, with the 4-L said to stand for "lazy loggers and loafing lumbermen." Yet the 4-L achieved more than improved production by gaining better camp and working conditions and in laying the foundations for the eight-hour day and the 48-hour week, with time and a half for overtime.

Disque brought in thousands of enlisted men, as members of the famed Spruce Production Division. Many of the men, wearing combat hats in the woods, knew nothing about logging. Moreover, morale among the soldiers was low, since they were stuck in the middle of nowhere, a long way from the fighting front, although they quickly learned that logging was equally dangerous. Even if true, there was no glory in it; a fellow couldn't become a hero or get a medal out here if he got clobbered by a log.

Disque administered the strange lumberjack army from Fort Vancouver, the historic Columbia River post near where the region's first sawmill came into being. Loggers worried that the troops were coming to maintain order, or to replace civilian lumberjacks. That wasn't the purpose; they were in the woods to offset a critical labor shortage. Many lumberjacks were drifting to the shipyards and other war industries where the wages were better. The greenhorns were sometimes more hindrance than help, for many troops hadn't been around timber, let alone seen the whoppers common to the Pacific Northwest. They had to be trained and also convinced that the job they were doing 6,000 miles from the war front was vitally important to the Allied cause, that it wasn't "inferior soldiering." It was a hard thing, for how does a fellow explain this to his girl and his family back home?

"These soldiers in the silent woods . . . will have a splendid part in the victory when it comes," reassured the *4-L Monthly Bulletin*. "And it is coming—with the toot of the spruce locomotive and the crash of falling trees and the shriek of the saws and the long trains of clean, clear spruce that builds the battle fleets of the air. . . ."

Such back-slapping pep talks, either verbal or in print, were needed to buck up the troops. Disque had 5,000 men in combat hats in the Big Woods by February 1918 and a peak force of 30,000, bunked at encampments established in the woods and near the mills. A large manufacturing plant was built at Fort Vancouver. The spruce was harvested, broken into cants at nearby sawmills, and then shipped to Vancouver or other specialty plants that could handle government specifications.

There was a frantic need for more logging railroads to move the big stuff. Ten thousand troops were assigned to laying 13 new roads, four of them permanent, along the Northwest coast from the Olympic Peninsula, at Grays Harbor, Willapa Bay, Port Angeles, and in southern Oregon. The Spruce Division assigned 4,200 doughboys and civilians, using 600 horses, in the building of the southern Oregon railroad. On the Olympic Peninsula, the need was urgent; men labored day and night to lay 45 miles of mainline and 20 miles of spur trackage. The pace was two miles a day at an average cost of $30,000 per mile against the usual low of $750 per mile, and running as high as $80,000. They got the job done, but not a single log was hauled over the Olympic Peninsula route before the Armistice.

Still, there wasn't enough quality spruce, so the Spruce Division began accepting Douglas fir. Officials were amazed to find that the Douglas fir was just as good. By the fall of 1918, the ratio of spruce and fir was running three to two—for every 30,000,000 board feet of spruce, 20,000,000 was cut in fir.

There was another surprising result. The war effort paved the way for the reforms in the woods, for Army regulations wouldn't allow the troops to live under the conditions which had long existed with the loggers. Camps assigned to the troops had to meet military standards, and thus other logging camps found it necessary to follow suit. Colonel Disque, armed with a vote of confidence by the employers, issued regulations setting standards for the woods, including a uniform charge of $7.35 per week for board and where there was housing furnished; the fee also must include clean bedding—sheets, pillows and pillowcases, mattresses, blankets—but with an extra $1.00 charged for weekly changes of sheets and pillowcases.

It all didn't sound much like the traditional logging camp, but the men welcomed the comforts, for attitudes were changing rapidly. But the stickiest issue of all remained the eight-hour day. After deadlocking with the workers, the employers passed the buck to Disque, at the same time accusing him stormily of trying to cram the shorter day down their throats. They hated it, especially since the IWW had promoted it. But Disque contended that the eight-hour day was desirable, since it would move this freewheeling industry ahead. The public favored the idea, and much of the nation's industry had already adopted it. Therefore, he ordered the eight-hour day, 48-hour week, and time and a half for overtime.

The war situation and public opinion left the operators little choice but to accept it, although some operators growled that once peace returned and the lumbermen were again their own bosses, they would go back to the longer schedule. It was wishful thinking; the workers were jubilant, and the unions immediately claimed credit for gaining this better work schedule at long last. If the operators didn't realize, it was almost a certainty that there would be no retrenchment once the fighting was over. The operators grew increasingly unhappy with Disque, although not forgetting the role played by the IWW. They could only express the hope that, as George Long stated it, the workers would now settle down to the business at hand and make "a more honest effort" to win the war.

When the Armistice was signed, the operators began backing away from the 4-L, since it had been a wartime device instigated by Disque. Its membership dropped from 80,000 to 8,777 by 1921. The 4-L was caught in a crossfire, for rival labor organizations brought withheld antagonism into the open. Even so, the 4-L stayed alive through the twenties and early thirties, often in direct conflict to other unions but never again to enjoy the influence it held during the war under Colonel Disque.

Bitterness against the Wobblies, who experienced a postwar revival, reached a new tempo, especially with the Russian revolution and since now Bill Haywood and company were campaigning for amnesty for "all political prisoners." The score wasn't settled yet by the lumber interests, as wartime emotions overflowed into postwar times.

The Washington lumber town of Centralia, some 50 miles east of Grays Harbor, stood in the heart of the big timber country. Local feeling against the Wobblies had been hardline for years and, in 1916, citizens ran a bunch of them out of town. But the following year the IWW boldly opened a hall in Centralia, purportedly one of only two within the state. Centralia citizens, among them many veterans and most all connected in some manner with the industry, wanted to wipe out that hall. During a Red Cross parade in May 1918, a group of "businessmen" raided the hall, nearly destroying the building. But the Wobs patched things up and stayed.

On the first Armistice Day after the war (November 11, 1919) a big parade was staged in Centralia sponsored by the American Legion. The Wobblies suspected there might be trouble. The parade was either scheduled to pass right by the hall, or the route was changed at the last moment. As with many such affairs, the truth of what happened was never learned, although volumes of words have been written from various points of view. There were armed men waiting within the building. As the parade went by a second time, the local unit of the Legion halted before the hall, either by design or because of a stall in the progress of the parade. Suddenly there were shots and the hall was rushed, or maybe it was vice versa. Men were wounded, the hall and its contents destroyed, and alleged Wobblies within the building placed under arrest.

A war veteran, Wesley Everest, lunged through the rear exit, firing as he ran. Legionaires were in hot pursuit. They caught Everest near the river, but not before he'd shot their leader. Overpowered and beaten, he was almost lynched on the spot, then hauled off to jail with a strap around his neck.

That evening, all of Centralia's lights suddenly went out. A band of men moved through the shadows toward the jail. The next morning Everest's still form was found hanging from a bridge. The execution was brutal; the man was castrated and riddled with bullets while he was hanging. No one dared cut him down. The body dangled there all that day and into the next before someone severed the rope, allowing the body to drop into shallow water, where it remained several hours before being taken to the jail to be displayed before other prisoners. Undertakers refused to handle the body. That night, four IWW prisoners were forced to bury the victim under armed guard.

Feelings ran to the extreme throughout the big timber country, to the point of hysteria. The American Legion, urged by the strong timber influence, took over law enforcement. The territory became an armed camp. Even the editor, president, and secretary of the board of directors of the Seattle *Union Record* were arrested on charges of publishing an unpatriotic editorial in violation of the Espionage Act.

Ten men were brought to trial for murder. The trial was moved to Montesano, but feelings were as heated there as at Centralia. A second change of venue was denied. Troops were stationed at the courthouse, and Legionaires were paid four dollars a day by the Centralia post to sit in uniform in the courtroom, purportedly an influence on the jury. The court refused to accept the jury's first verdict. On the second, seven were found guilty of second-degree murder, two not guilty, and one judged insane.

The Centralia Massacre, as it was called, would take a long time forgetting and for the timber industry to live down. The tragedy plus the anti-red hysteria of the twenties brought the IWW under attack more than ever before. Since the industry was the dominating power of the Northwest, the people were more determined than ever to run out the tough-talking, troublemaking Wobblies.

Yet its card-carrying members remained fiercely loyal to the IWW as "the laboring man's union." It was a die-hard organization even with a declining membership, continuing to revive itself from time to time and keeping a Seattle office open into the midcentury. In 1972, a bearded organizer, Frank Cedervall from the IWW Chicago office, was touring the Northwest, trying to revive the Wobbly movement, and speaking—of all places—on college campuses. By this year, much of what the Wobblies advocated in reforms was an accepted part of the American scene, and far greater benefits than old loggers would have rated as pie in the sky. Cedervall was a ghost from the distant past, the extension of a bygone age, yet he contended that many things for which the Wobblies stood are just as pertinent today. He found much interest among the dissatisfied young, the graduate students, and people with degrees unable to find work. The IWW, he said, today is no longer a movement but an ideological organization, its 5,000 members including a student here, a professor there, and even an occasional laborer.

"Do you know there are 60,000,000 unorganized workers in this country?" asked Cedervall. "Out of 60,000,000 we can surely organize a couple of hundred thousand. And that's all we need!"

The Wobblies died hard. Or maybe they never did.

Chapter VIII

Conflagration . . .

Forest fire and a smoke-laden atmosphere are of almost yearly occurrence in this region, but seldom, if ever, has there been anything of the kind equal to what is now making life unpleasant. . . .

The Oregonian
September 1902

From early summer until the autumn rains of late September or October, and into most recent times, the Douglas fir region was struck annually by raging forest fires. It was part of the accepted routine, and it was a most unusual year when these sorry holocausts were at a very minimum.

The crucial months were July and August, when the dry east wind blowing from the rangelands of eastern Washington and Oregon sucked the protective moisture from the Big Woods, leaving them massive tinderboxes, as explosive as a munitions dump, extremely vulnerable to any carelessly tossed match or cigarette, an unguarded campfire, a piece of glass catching the sun's rays, or a strike of lightning. Yet in the lawless years, one of the prime causes was the logging industry itself, which you would ordinarily believe would go to an extreme to protect the green bonanza that sustained its life. But the trees were sacrificial, and while some precautions were taken, the average lumberjack and boss logger believed that you couldn't keep fire out of the woods, especially in this mechanized age. The sparks from a passing locomotive, from a donkey engine, or the friction of one log passing over another could in seconds cause all hell to break loose in an out-of-control rampage through thousands of acres of fine virgin growth. The loss was tremendous, with the Green Desert turned into a charred, blackened, desolate wasteland where the trees wouldn't come back for at least a century.

Still, fire was accepted by most everyone in the logging-lumbering community as one of the risks; there were few men indeed like Colonel Bill Greeley, who hated fire with a passion and genuinely believed that it could be licked. Fire was a necessary evil; the loss of a fine forest was a sorry thing, to be sure, especially for the operator who shelled out his dough for the stumpage, and the men who were out of jobs, but there were always other forests and other rich stands stretching out of sight across the rolling hills, so that the operator might obtain a new deck and deal again.

Fire belonged to the recklessness that characterized the industry and followed the loggers like a plague clear across the continent. In the timber belts of Wisconsin, Michigan, and Minnesota, there was great disaster and destruction, the loss of many lives, settlers driven from their homes, a shattering of the economy, and the loss of rich timber stands that could never be grown again. Peshtigo . . . Hinckley . . . Cloquet . . . these names were not only sorry milestones in the ravaging of a resource, but also played a major role in shaping the cut-out-and-get-out philosophy of the industry itself, with timber looked upon as an asset for quick liquidation instead of a resource that could make the industry a permanent part of the American scene.

Because of the rain-washed Pacific Northwest shore, the moist climate, and the thick evergreenery that often wouldn't produce a warming fire on a chilly morning, the timbermen felt more secure against the blackening ravages that haunted other fire-lashed forests. But it simply wasn't true. During the summer dry seasons, and even in early spring or late fall, the Northwest woodlands could rapidly reach the blow-up stage, where a single small spark, smoldering for days or weeks in the crack of a fallen log or amid some dry brush, exploded suddenly into a holocaust.

Early loggers also took some consolation in the evidence that mighty fires had destroyed lush forests of this bonanza territory long before their arrival, from lightning, the friction of branches rubbing together in a wind, or the Indians, who were careless with their campfires. The Indians also fired the forests with expressed purpose to clean out the dense timber and underbrush, thus opening the country for game and improving the hunting. This held true after the white man began burning up the woods. The years following the several Tillamook disasters, the Big Burn became one of the finest deer-hunting regions in the nation.

Throughout the Northwest are still evidences of great forest fires of centuries past. One of the largest, and oldest, stretched along the Washington coast from Willapa Bay deep into Humptulips country, the Promised Land area, and beyond the Ozette Lake region. Evidence showed that this ancient fire happened at least 500 years ago, for the stark forms of huge cedar trees reach ghostlike for the sky, some dating back to the beginning of the Christian age. A 500-year-old "young cedar," born after the fire, was discovered growing from the remains of an old cedar, which fell in that conflagration and showed rings placing its age at 1,500 years. That plus the 500 years of its offspring set its beginning in the time of Christ.

Those who studied the old Burn were unsure whether the disaster was a single fire or a chain of fires, all burning simultaneously or in rapid succession. It is a moot point; the damage was done, and its cause will undoubtedly never be known. But from this calamity, which turned the country desolate of what must have been an awesome forest of cedar, emerged the Douglas fir as the region's dominating species. Its seeds blew on the Grays Harbor winds across the land, and the healthy seedlings sprang up to grow thick and tall and untouched for several centuries, becoming timber's Comstock Lode, which attracted and left speechless the first loggers and lumbermen. As one way of looking at it, the fire had been a good thing, a prime example of how in this timber-growing region, nature heals itself if given the time and the right ingredients, and if free of the impatience of man.

Settlers fired the forests and were happy to see the damned things burn. The result was opening the country for farming and grazing, although much of the timber land was an agricultural mistake. Stumpland farms sprouted on logged-over lands, but they were normally small and unproductive, for the forest land just wasn't suited to this kind of activity. But fortunes in timber were going up in flames year after year, timber not only for now but for the future, and that meant sprawling areas that were idle and valueless. More and more people realized it was not a minor problem but a major tragedy, and that sooner or later everyone —the industry included—would need to deal with fire in a hard-headed manner. With so many independents working in the woods and the public using the forests for hunting, fishing, camping, and general recreation, it appeared only a distant dream that anything could ever be done. Then the late summer of 1902 became the first turning point, for in that year all the Douglas fir belt was ablaze.

It is called the "Yacolt Burn," but it was far more than a single fire. Timber fires were raging from the Canadian border at Lyman, Washington, to the Oregon-California line, and from the western slopes of the Cascades to the sea near Grays Harbor and Astoria. More than timber was lost this time: Mills, towns, logging camps, equipment, and people's lives were consumed by the flames. The dangers were real and ominous as heavy smoke palls choked the country, ashes fell on the streets of the cities, and fire surrounded the villages and smaller towns that belonged to the timber world. Portland, Tacoma, and Olympia were shrouded in smoke. In Astoria, the setting sun through the thick layers of smoke turned the sky a "yellow green," while upriver places like St. Helens and Scappoose were dark at noon, the railroads keeping their stations lit and the trains creeping along, attempting to maintain their schedules in this choking shroud. The Pacific Northwest had seen nothing like it, and there was little that could be done to control the fires until the humidity rose and the rains came to quiet the flames.

A hundred villages and towns were threatened on September 12, called Black Friday. In the bleak noontime darkness, many people truly thought the world was coming to an end—that the sun was being permanently blotted out. One of the hardest-hit areas surrounded Grays Harbor. Many huge fires were burning through good timber and slashings. The heavy smoke pall was driven west toward the sea, and when it hit the marine air, it settled down over the lumber centers. Sawmills attempted to run at 7 A.M., but two hours later closed down because of the poor visibility. Lanterns were carried along the streets, and at mid-morning steamers were forced to resort to searchlights. Crowning fires in the distance were frightening sights. While Aberdeen and Hoquiam weren't threatened seriously, the flames crept to the edge of the cities, and that was too close for comfort. Mayor W. W. Anstie of Aberdeen tried to calm his people, reassuring them that there was no cause for alarm, even though there was fire in most every direction.

"Everything goes down before the fire," observed the *Aberdeen Daily World*.

"Mills are destroyed, houses vanish in smoke, and the earnings of a lifetime are destroyed. People who were happy and prosperous yesterday are homeless today."

The town of Elma, near where this particular blaze began, was now threatened. Elma's saw and shingle mills were leveled, as were two logging camps owned by the Polson Brothers and the Elma water works. The dam on the Hoquiam River and the mill at White Star, with 1,000,000 board feet of lumber in the yards, were wiped out by the holocaust, as was the village of New London outside Hoquiam, a camp occupied by Norwegians.

"No one can conceive the scene of desolation who has not visited it," commented Mayor J. R. O'Donnell of Elma. "From Summit nearly to the Satsop, a strip 13 miles long and from one to two miles wide is a mass of charred and smoking ruins." *Charred and smoking ruins . . .* the words would be used thousands of times in succeeding decades to describe the aftermath of Northwest forest infernos. There is nothing more desolate than the black, smoldering remains of a once-beautiful forest, save perhaps the cities of Japan following the atomic explosions. Nothing stirs, nothing lives, and all hope seems gone.

Destruction and terror were repeated again and again throughout Washington and Oregon as the season wore on, until the accounts took on a sameness, with only names and places changed. Penniless homesteaders striving to carve out a simple living from the raw wilderness were forced to flee for their lives, losing what few possessions they had accumulated. Sixty were homeless when the village of Springwater near Oregon City was wiped out. There was a big loss in roasted cattle, horses, and hogs, but no mention was made of the tremendous loss of wildlife, for like the timber, this was an overabundant resource considered of little value in those times. The entire region remained a tinderbox. The huge sawmill at Bridal Veil was leveled by flames starting from a railroad locomotive. Two miles away, the little town of Palmer was engulfed in fire out of control in a high east wind roaring down the Columbia Gorge. A salesman from San Francisco had a harrowing time, fleeing ahead of fast-spreading flames in his buggy, with the heat so intense as to be almost unbearable. Everett was now cut off from the outside, for communication overheads were being destroyed everywhere. Centralia was choked in smoke, with ashes and cinders falling into the streets, while the smoke became so heavy across the treacherous Columbia bar that ship movements were halted. Still, the fall regatta of the plucky Portland Rowing Club was staged in a smoke-swirling haze.

Over 110 fires were all roaring out of control in what was described by the *Seattle Times* as "the worst in the annals of the great Pacific Northwest," considering what unhappy twist of fate had bunched all the timber region's annual forest fires into one angry, tragic spectacle. Still the flames roared on, the smoke clouds billowing thousands of feet into what had been a deep blue sky, appearing as atomic clouds would to another generation, the towering yellow flames crackling through the woods and leaping for many miles over the treetops, while inner winds within the holocausts built into twisting, tearing tornados that picked up great trees and logs, hurling them across deep, flaming canyons resembling the bowels of hell itself. Still, the worst was coming in southern Washington, north of the Columbia River and east of Woodland, in Clark County, in what made the summer and fall of 1902 a time to remember.

Yacolt was a tiny hamlet of some 15 buildings and shacks, a division point for the new Portland, Vancouver, & Yakima Railroad. When fire broke loose from

16. When the steam donkey invaded the Big Woods, the bull teams were retired, and at long last, the loggers had a form of energy beyond their own muscle and that of animals. —*Oregon Historical Society.* (Chapters II and III)

17. Horsepower and wooden rails had their place in the woods, such as in Tillamook County, where the greatest forest fire of them all would one day occur. Wooden rails were forerunners of things to come. Loggers were slow to accept change and often made do with what was available and within reach of their pocketbooks. —*Oregon Historical Society.* (Chapter III)

18. Horse logging is still used (note hard hat) under certain conditions in the Big Woods, especially where the "ecology" is touchy, such as in watersheds and reservoirs, or for thinning on tree farms. —*Author's collection.*

19. Colorful log drives down the many rivers were another primitive method of getting the sticks to the mills. This struggle was on the Trask River of Oregon about 1915, 18 years before the great Tillamook disaster. The Northwest's last major log drive occurred in 1971 along Idaho's Clearwater River. —*Oregon Historical Society.*

20. It took muscle, skill, bravado, and no small amount of cursing to move heavy timbers like this fat spruce log bucked near Seaside, Oregon. —*Oregon Historical Society.*

21. When railroad logging came to the Big Woods, following invention of the Shay locomotive, lumberjack engineers built towering trestles to the high places. They were remarkable engineering feats, but often scared the daylights out of the loggers. —*Oregon Historical Society.* (Chapter III)

22. The pokey lokies—geared Shays, Heislers, Climaxes—were said to run as much on the ground as on the rails, from high ridges to deep canyons to haul out the fat timber. —*Oregon Historical Society*. (Chapter III)

23. The bigger logging outfits had fleets of locomotives like this one of Yeon & Pelton, representing a sizable investment. —*Oregon Historical Society*. (Chapter III)

24. Whether by skid road, flume, or railroad, all the sticks headed down to splash at the landings on the rivers and bays, and then to the sawmills. —*Oregon Historical Society.* (Chapter III)

25. The marvels of the block and tackle, using huge blocks and thick cable, coupled to a donkey engine, wrestled the whopping big ones from the woods. This was at Valsetz in the Oregon Coast Range. —*Author's collection.*

26. Traditional sawmills were rustic, utilitarian, noisy, and poorly lighted. At the head rig, the head sawyer decided how to cut up each log to get the most from it. He became a master craftsman. Today, computers make the decisions. —*Author's collection.*

27. Loggers long dreamed of a skyhook, to "fly" the timber from the deep canyons and high ridges to the landings. This contraption was an early version. —*Oregon Historical Society.*

28. The wigwam burner solved the problem of sawmills being buried by their own debris and became a landmark of lumbering. It also created a blackout crisis following Pearl Harbor when the Big Woods was lighted up like a Christmas tree. Today, the environmental movement and a sharp decline in waste are phasing out the landmark burners. —*Western Wood Products*. (Chapter III)

29. Forest fires were long an accepted part of mining the green gold. Here, a logger gazes down upon the last of the great Tillamook fires in 1951. —*Ellis Lucia photo*. (Chapters VIII and IX)

30. In 1933 holocausts of the Big Woods reached their highest pinnacle in the great Tillamook disaster. The fire literally blew up like an H-bomb on August 24, rampaging over 220,000 acres of fine timber in 20 hours and creating a mushroom cloud 40,000 feet above the disaster. —*Author's collection*. (Chapter IX)

rancher slashings, bearing down on the town, the people fled to the creek, where they spent a frightening night. The land of the Lewis River was described by a newspaper's country correspondent as "a hot and silent valley of death." The fire moved at high speed. Two families, celebrating the arrival of an uncle from Kansas, were traveling to Trout Lake for a picnic when flames suddenly rampaged down upon their lumbering wagon. The five adults and six children, one a baby, attempted to outrun the speeding flames, but got only a few yards, and the freed team not much farther. For a month no one knew what happened to the families, although all feared the worst. The remains of the wagon and occupants weren't discovered until the embers cooled on the autumn rains. As is often the case, the families were ironically only about 100 yards from a creek where they might have found reasonable safety.

The Yacolt became a killer as others died—a woman who was attempting to save her Singer sewing machine, her most cherished possession, and 200 jars of fruit; a Star Route mailman whose charred body was discovered against a log, clothing and even the skin burned away; a woman in the Dole district and her three children who sought refuge in the cellar of their house, which caught fire and then tumbled inward, turning that cool cellar into a blast furnace; several other families with young children who tried to flee but never made it. In all, 35 died in the Yacolt fire, but in another strange twist, the hamlet for which it was named escaped with only damage to house paint, which was blistered and scorched by the searing heat.

The horrors of Yacolt, and all the fires of that summer, weren't easily dismissed as just another bad season. The memory would linger long, becoming an early focal point on this entire matter of forest fires, for both lumbermen and the public began realizing that this waste of resources and lives couldn't continue forever. When all was totaled, that flaming September consumed some 12,000,-000,000 board feet and a total of 239,000 acres, at a monetary loss of $13,000,-000. Added to this was the heavy toll of lives. The timber loss was three times the region's annual cut. In the Columbia district alone, 1,500,000,000 board feet were destroyed. There was also the loss of many small trees, the hope for the future, if there were to be any future in the forests.

Still, it is an unhappy fact that in these United States, it takes a sobering event such as this, accompanied often by death, vast destruction, and financial loss, to jolt people into seeking another way of doing things. The impact of the conflagrations of 1902 stirred the public and the lumber leaders toward a system of better protection for the forests. Even so, it would require half a century, at least three other mighty conflagrations, and the arrival of an amazing forester, Colonel William B. Greeley, before the fire problem could be licked; and all the while there were thousands who said it was impossible to achieve, and still harvest trees. But they began listening to the visionaries like Everett Griggs and George S. Long, the lanky boss lumberjack who helped pioneer forestry for the Weyerhaeuser Timber Company after they brought him from the Lake states to administer their Pacific Northwest holdings. Men such as these, before the gatherings of lumbermen and loggers, contended that wisdom must come to the woods, that they couldn't continue forever operating as in the past, and that they had a responsibility to the public and to future generations. Long warned:

"Every additional year means an increase of the fire hazard, and if the logger does not adopt precautions to reduce this fire hazard to a minimum . . . the public is going to remain in a hostile frame of mind and the public is going to be unjust to him in its enactments on the statute books. The logging operations in Washington and Oregon have been largely in regions which have been remote. If history repeats itself as to forest fires, it will not be improbable that some day we will not only see vast areas of timber lands wiped out by forest fires, but the homes of settlers and towns, with a deplorable loss of life. Such was the history of forest fires in Wisconsin, as for instance, the Peshtigo fire where the loss of life was over 1,000 people, and the Hinckley fire where over 150 people burned. . . ."

It had already happened at Yacolt, although still on a small scale, and there would certainly be other tragedies. The region's leading timber loss was from fire, often well beyond the annual cut and the tremendous waste. In the public mind, the timber operator was to blame, and often he was; but on the other hand, no wise operator with an investment in forest acres wanted them destroyed any more than a farmer would burn his wheat crop. The crux of the problem lay with the freewheeling boundaries of the big industry itself and its many mavericks, the lack of laws and adequate controls and standards for preventing fire, plus the fact that the timber lands were open to all comers. The public—hunters, anglers, picnickers, hikers, and vagabonds—was free to invade the forests at most any point, since most of the Northwest was still undeveloped frontier. Its people held a don't-fence-me-in independence that remains a charm of the Northwest today and at that time characterized much of the law of the land. But the public also was more careless with fire than were most loggers. During the wet seasons, it didn't matter much. Loggers, woodsmen, sportsmen, and the public at large built warming fires and tossed lighted matches and cigarettes into the brush, for the great woods were too water-soaked to burn. The habit carried over into the dry seasons. People couldn't believe the woods would dry out so rapidly to a stage where a small fire would take off in seconds and travel for miles. Once when I was a correspondent for the Associated Press in timber country, I telephoned the Portland bureau to report a dangerous early-spring fern fire, after a very wet season.

"Hell," said the man on the other end of the horn, "you must be drunk. There can't be any fires this early."

I had to explain in detail how rapidly the fern and brush dry, once the sun comes out, and how small trees are destroyed in such fires.

The fires of 1902 provided a graphic illustration, originating not just from logging operations, but also from a broad spectrum of smaller blazes, including the warming fires of lumberjacks and campers in the spring, those of late-summer huckleberry pickers, and one of a small boy firing a yellowjacket nest. Then in September, all the ingredients combined, and the dynamite cap that ignited it was a humidity plunging nearly out of sight. There needed to be an attempt to lessen the fire causes, and the Yacolt provided the impetus. Fire patrols and protection organizations were created by timber owners who refused to wait for slow state legislatures and wished to guard their holdings. The aim was to reduce, if not eliminate, the losses from forest fires. Owners often pooled their lands and resources, including equipment and manpower, financing the protection costs pro rata, based on the acreage of each owner.

Then the Washington legislature surprised everyone. The year following the Yacolt, it created a state forester and a chief fire warden, and two years later (the next legislative session) established a state board of forest commissioners, with the power to appoint a state fire warden and deputy foresters. Oregon took a little longer, establishing its first state board of forestry in 1907, but it gave it little authority and only $500 to operate for two years. These steps were minor exercises in futility and of small effectiveness, since the state authorities were operating without forceful laws and under the control of the industry, whose tough membership didn't want very many restrictions to their methods of operation. For that matter, neither did the public wish too many laws telling them what to do in the woods. The industry attitude remained that "I'll log where and when and how I damn please." However, the private-protection idea spread through the two states and into Idaho and California during the decade to 1912, and slowly better fire codes were developed, patrol laws were passed, and better fire-fighting techniques were pioneered.

The fire-protection associations took to the legislatures such suggestions as compulsory patrols during dry weather, required fire-fighting equipment at all camps, closed slash-burning seasons, forced shutdowns during dangerous humidity conditions, and compulsory snag falling and slash disposal. The associations also gave timber operators the opportunity to work in harmony on what was considered a common problem, which years later caused Bill Greeley to comment that "it was here that industrial forestry began in the West."

Then the awesome 1910 Idaho-Montana holocaust broke loose, seemingly a repeat of 1902, with 1,736 fires burning in national forests alone, and more than 3,000,000 acres destroyed. At least 85 lives were lost, most of them fire fighters on the lines, while the timber losses again ran to billions of board feet. The ghastly affair made national headlines, which aroused the public and in turn the Congress, and nearly cost the career of the young Bill Greeley on his first real assignment as a ranger on the Idaho-Montana blow-up. It might have discouraged him to quit, but instead it created a determination within to do something about forest fires.

The following year the Weeks Act was passed, giving federal assistance to state and private timber owners to maintain protection organizations and private fire-fighting corps, and encouraging states to set up forestry departments, already an established fact in the Northwest. It was another short step in the right direction, but it didn't lick the fire problem by any means, although fire fighters were now attempting to control the outbreaks rather than taking the view of "let it burn 'til it rains." Still, both the laws and the attitude were a long reach from making the general public understand its own major responsibility, and also from convincing many logging and lumber operators. The laws simply weren't tough enough. There was also the character of the fire-protection associations, which were "private" and therefore were sometimes reluctant to shut down operations in dangerous weather because the wardens must answer to the operators who contributed yearly to their salaries and activities. So year after year the woods went on burning, throughout the teens and the Roaring Twenties. Then in the 1930s, two memorable explosions in the coastal timber country of Oregon, three years apart, renewed the outcries for stronger laws and tighter restrictions on the industry to keep the nation's remaining forests from being completely destroyed. Both of these great holocausts broke loose only a few years after Bill Greeley came to the Northwest hoping to cure what he described as a "sick industry," still hating fire, and still convinced that the only solution to forestry for the future had to begin with a workable system of controlling and reducing forest fires. Then came those two explosions, costing many lives and the loss of millions of dollars to Oregon—the first, one of the greatest forest fires of all time; and the second, fulfilling George Long's warning that eventually towns would be destroyed if the old ways continued to prevail.

Chapter IX

The Great Tillamook Fire

A wind of hurricane force and shifting propensities is howling over the mountain tops today, driving the flames first this way, then that. The 1,500 men on the lines are as important as a fly in a whirlpool. They are working hard, patiently and without complaint, but fire lines are jumped before they are finished and backfiring almost amounts to starting new forest fires. The newest jump of the fire seems likely to extend to the east–west area almost the full width of the mountains, with so much smoke over the hills that no one knows the exact north and south dimensions. The Wilson River fire, wardens agreed, was a solid mass of flame beneath the heavy pall of smoke. . . .

MARY L. ROBERTS
Correspondent, *Oregon Journal*
Tillamook Fire, 1933

West of Portland in that jutting extension of Oregon bounded by the sea and the final northward swing of the mighty Columbia River lay one of the finest and most entrancing timber areas of them all. It ranked on a par with the old growth lands of Grays Harbor and Puget Sound—thick, beautiful, big woods, seemingly destined to supply the needs of the American people for a good many generations. And like Grays Harbor country, the sprawling and rugged territory was a mist-washed natural paradise for growing stately timber.

"From the summit of the Coast Range Mountains to the tidewater line, a distance east and west of approximately 35 miles, it is simply one vast and dense forest," wrote a newspaper reporter traveling through the woods in 1902, the year of Yacolt. "Through this vast region, there are some places of scanty growth, with some areas which in times past have been burned over and destroyed, but broadly speaking, it is a virgin and all but unequaled forest district, in which the fir predominates, with a sprinkling of cedar, and toward the coast, a great wide area of spruce. And everywhere, it is a forest area of the giant breed, with trees ranging from eight to 30 feet in circumference and reading upward from 150 to 300 feet. There are men who will undertake to say how many multiplied million feet of merchantable timber there are in this great forest, but I will not give their figures, because I have no faith even in their approximation of accuracy."

There had been bad fires in this region, touched off by lightning, the Indians, loggers, and outdoorsmen, and even after many years of rapid regrowth, the scars lay upon the land. But in 1933 all of this giant woods was doomed in one of the greatest forest fires of all time, and surely the most memorable of this century.

The woods had been bone dry for two months, making logging touch-and-go, for the humidity was often very low. Logging operators were using caution, hoot-owling from dawn when the humidity was up and then closing down early in the day, refusing to risk a gamble with the gods of flame. Yet foresters and timbermen alike feared for the woods of the Northwest, for this was the Great Depression, when greedy and shortsighted men created jobs by setting fire to the forests and then signing on for pay to fight them.

This particular day was extremely hot and dry in the Tillamook country, with the humidity plunging steadily downward. Operations were shutting down early, the lumberjacks drifting back to camp and village to ride out the heat in their bunks. There were no shortwave radios or walkie-talkies to send out quick warnings to woods operations. "Runners" were dispatched to the more isolated gyppo shows. But laws were still mighty loose; the responsibility was left largely to the judgment of the timber operator and his crews, who often went by "feel." The hygrometer was a comparatively new instrument; not every operation had one, and many loggers didn't understand what a dropping humidity meant to the woods.

About two miles above Glenwood, the outfit of Elmer Lyda of Gales Creek, whose father had harvested this country with bull teams in pioneer times, was punching logs for the Glenwood Lumber Company on a steep hillside above the railroad in timber and logged-over land of Crossett Western. They'd been there since spring, winning a highly coveted bid, and considered themselves fortunate to be working, since so many operations were down and men out of jobs. Furthermore, the timber company had extended its railroad deeper into the trees with a series of switchbacks. If Lyda Logging Company did a good job here, there was promise of additional contracts along that railroad where enough timber stood "to last my dad the rest of his life," son Bill Lyda said years later. "It looked like we got something started that was really good."

Elmer Lyda was no inexperienced timber beast; he'd been logging for about 10 years and so had his son, Bill, who began working in the woods at 14 and was now 23 and a partner in the company. With Elmer's daughter keeping the books and his son-in-law also on the show, it was very much a family affair.

The Lydas supplied logs to the nearby sawmill by rail at a rate of 75,000 board feet per shift, for $16 a thousand. This kept them under constant pressure, for the mill gobbled timber like a horse eating hay. Sometimes they worked Sundays at rerigging to be ready for Monday morning. They had a top-notch hand-picked crew of about a dozen men, for in this sorry year of the Great Depression, you could hire the best in the trade for $3.50 a day; they were glad to get the work, since hardly anything was moving, and good men were doing odd jobs for as little as $.20 an hour. But there was happy news this week, for the price of logs was being increased $4.00. It seemed that things would be looking up for the balance of the year. All told, a silver lining appeared to be hovering over the Lyda family, although there had been some bitterness, particularly from one other operator, when they won this contract with Crossett Western.

That was how things stood on the morning of August 14, 1933. Four decades later, nearing the fortieth anniversary of the Great Tillamook Fire, William H. Lyda, the son who was on the scene, gave his version of what happened on that fateful day to me in a tape-recorded interview at his home. Over the years since the big fire, hundreds of published articles in newspapers, magazines, and books fixed blame for the holocaust on the Lyda family, charging negligence, recklessness, and greed in failing to heed the warnings and instead yarding "one more log"—the log that blew up everything. Since the accepted version, now a legend, has been related countless times, it seems worthwhile to tell the other side from Lyda's point of view, a long-overdue rebuttal, including the overlooked detail of a mysterious second fire that appears to have been raging ahead of the Lyda outbreak, deep in the forests of the Wilson River country.

"We never denied that we had a fire," Lyda said somewhat bitterly. "But everybody else has denied that there was another one 5¼ miles over a 2,600-foot elevation that started maybe four or five hours before our fire had started. They've never admitted that."

Because of the heat and fire danger, the Lydas were hoot-owling, beginning work as soon as it was light. They were very much aware of the dangers and certainly didn't want a bad fire, since the prospects of this contract would ride out the Depression for them. Under the Crossett Western agreement, they took their shutdown orders from the timber company manager, who had an office two miles away at Glenwood. He had a psychrometer; they didn't. When the psychrometer dropped to 35 per cent, he was to come by railroad speeder to the landing and shut down the steam donkey, which was located right at the track. Its shrill whistle would sound the warning to the loggers on the hillside, where there were two gas-driven machines. There was a leeway of five hygrometer points on the hill, but when the steamer whistled off, they would kill everything immediately.

At midmorning, the crew knocked off for lunch. The men remarked that it was another scorcher and getting pretty dry. Elmer Lyda went down to the landing office, while Bill hiked out on the hill, near the towering 211-foot spar tree, to look things over from that vantage point. They'd been using a rub tree to bend the skyline to avoid a hangup, but that morning Lyda told the men they'd "better take that rub out of there 'cause it was getting pretty hot." It meant rerigging to get rid of the skyline. Now they were fighting the hangup that the rub tree eliminated. Lyda was worried; one of the gasoline engines wasn't acting right, and he felt that something was generally wrong. It was terribly hot, and the humidity seemed very low.

Below Bill were four loggers, with his brother-in-law acting as foreman. Lyda called down, saying they'd better come in, for things looked critical. His brother-in-law said that they had four or five more logs to bring in, then they'd change the line to make ready for the next day. He also pointed out that the speeder hadn't arrived, so the humidity must not be down to deadline.

The men went back to work on those four or five logs. Lyda didn't remember which log was the fateful one, except that it wasn't the last—the "one more log" of legend—that hung up, rolled slightly, then was dragged across a cedar wind-fall near the yarder pole. Suddenly, at 12:26 P.M., with an explosive "puff," as though touched off in gasoline, flames shot out of the brush and debris *behind* the log's tail as it passed over the windfall. The published versions say the fire started when the two logs rubbed together.

Lyda maintains this isn't true because of the position where the fire began, although he concedes that it could have started by friction of the rigging or the collision of the two logs. However, in going over the ground, he discovered three flashlight lenses in the rubble that weren't there before and could well have started a tiny fire in the combination of sun, heat, and low humidity. The dragging of the log across that area might well have turned a smoldering tiny hot spot to instant flame. The operation wasn't using flashlights; how the lenses got there, nobody knew. But the Lydas long kept those lenses in their home, until it burned in 1955, and it is Bill's suspicion that sabotage and revenge were involved by the embittered operator whom the Lydas had outbid.

The loggers surged onto the rapidly spreading fire. Quickly the flames destroyed the remaining logs waiting to be hauled in. The alarm sounded down the hill and at the mill, and shortly crews from the saw and shingle mills joined the Lyda men until some forty hands were hitting the fire. They managed to trail around the upper side, but they couldn't save the gas engines. The flames spread in the opposite direction, toward the railroad landing and the creek.

Meanwhile, sawmillman Russell Raines telephoned Forest Grove for help. When he couldn't reach forest warden Cecil Kyle—he was up in the woods somewhere—Raines contacted Walter Vandervelden, the firewise chief of the town's volunteer department. Raines said he was sending his wife in with a pickup truck, and would Vandervelden go over to Kyle's house, which served as the little district forestry station, and help her load up whatever available equipment was there? Vandervelden agreed, and found Frank McCullough, Kyle's aide (also his brother-in-law) at Kyle's place. They loaded two pickups with marine pumps, hose, and other tools, and took time to change the oil in the district's truck, since it was so thick the engine would hardly turn over. Then Vandervelden climbed in beside McCullough for the 14-mile run to Glenwood, with Mrs. Raines out ahead of them.

Leaving town, Vandervelden spotted not one but two smokes.

"Frank," he yelled, "what the hell? You got two of 'em going?"

McCullough took a glance. "By gosh, it looks like it."

"Yeah," shouted Vandervelden, "it looks like the right-hand one over here is yours, but the other one looks like it's a little farther back on up."

The men continued to see the two fires for many miles, until the angle changed and the far one was no longer visible.

At Glenwood, they shifted the gear onto John Markee's motorized log hauler, which he used to bring the logs down to the sawmill, and then made the two-mile trip over the railroad to the Lyda landing. The loading took three quarters of an hour. By the time they reached the scene, the fire had spread down the hill to the landing and Vandervelden considered the situation hopeless, for there was so little equipment, they couldn't get at the fire, and it had already burned quite a wide area. The opposite logged-off hillside of slashing was being ignited by spot fires, having jumped the creek, and for a time Elmer and Bill Lyda and several other men were trapped, with fire surrounding them on three sides, to be rescued by the pumping equipment brought up by Vandervelden and McCullough.

Foresters converged on the scene. C. S. "Scotty" Scott, secretary of the Northwest Forest Protective Association, arrived from his Portland office and took command until state forester Lynn F. Cronnemiller got there from Salem the following day. Warden Kyle came in from somewhere, evidently having seen the smoke. Still, from the Lydas' point of view, there was much confusion as the foresters took over, feeling that they lost control of their crew at a time when they were planning to backfire and make a stand near the switchbacks. The loggers believed they had the fire nearly licked, that they could well bring it under control by end of the day. To add to the confusion, Elmer and Bill Lyda were ordered off the fire at a crucial time to tell about their insurance. And they were astounded when the Lyda men were pulled from the lines about four-thirty without explanation.

Warden Kyle put in a request for help from the CCC camp back in the forest, a force of some 200 boys, but they didn't hit the fire until next morning. By then it was burning in a railroad trestle and slashings from a five-year-old logging operation, and was beginning to crown.

"That's when it took off; that's when it went to the sky," said Lyda. "Not a few minutes after the fire started."

On the first afternoon, the fire moved over the ground and never crowned, he related. But a second fire of unknown origin was on the rampage nearly six miles to the west, and Lyda contends that the second fire was really the damaging one. Walt Vandervelden saw the second smoke, as did Frank McCullough and some fishermen, among them the Watrous brothers, who were well up in the woods. Yet the weight for the great Tillamook disaster fell on the Lydas, the contention being that on a high wind, fire brands were carried five to six miles over the ridge to ignite the Wilson River area—that *everything* spread from the Lyda fire. But fire chief Vandervelden, who served four decades maintaining one of the nation's finest volunteer departments and is considered a man of much integrity, said only a light wind of five or six miles an hour "that you could pace by walking" was blowing, not sufficient to carry sparks or brands a considerable distance.

Postfire hearings held in Portland under the auspices of the timber companies brought out facts about a "second fire," including the testimony of Walt Vandervelden. The hearings were private and without press coverage. The Lydas presented 24 affidavits supporting their contention of another fire raging farther west well ahead of their noontime outbreak, and that their fire didn't "jump" several miles beyond the ridge on that first afternoon. But in the end, the conclusion was drawn that the Lyda fire was to blame for what happened during the next two weeks of that terrible August. The legendary version of "one more log" was published again and again.

"It hurt the whole family," says Bill Lyda today, "and it will hurt us as long as we live." He feels that they were made the scapegoats, perhaps to cover up the cause of the other fire, which might have fixed the responsibility on big timber interests in the area, making them the prime targets for lawsuits and being charged with the loss and fire-fighting costs.

The Lydas were forced to take bankruptcy, and Bill Lyda, because of local animosity, moved to another part of the state, where he stayed in the logging business. His father gave it up, instead going to work for Walt Vandervelden and his brother in their machine shop, building logging equipment.

Things might have been different if the speeder had arrived to close them down. Lyda learned that the psychrometer read 14 per cent at twelve twenty-six, when the fire broke out.

"We should have been out of the woods at the time we ate dinner," he commented.

Why hadn't the hygrometer man come? Alf C. Johannesen, who was in charge for Crossett Western at the Glenwood Camp, noticed early that morning that the humidity was dropping fast. At ten o'clock, when it was still above deadline, he decided to close down Crossett Western's two logging operations, the Lydas and John Heisler's. He chose to inform Heisler first—"it was six of one and half a dozen of the other"—since he had to drive there by car and also had to pick up the mail at the post office. After stopping for the mail, he headed for the Timber road toward Strassel, six or seven miles away from the Lyda operation in the other direction. By the time he returned and reached Lyda's, the fire had broken out. He was about thirty minutes too late, and all he could do was get his pumper and start fighting.

Johannesen was aware, too, of that second fire, and testified about it at the hearings. But the second fire's origin became a mystery, since there were no logging operations in that wilderness country. However, people were scattered throughout the forest, including sportsmen and a 3-C camp. Lightning has also been mentioned as a possibility. What might have touched it off?

"I was told there was a big wingding party in there the day before," Lyda said. "And next morning a woman called down to Forest Grove, asking someone to come get her and the kids, as they were surrounded by fire."*

Nevertheless, the fires were on the make and soon would become a holocaust to end them all. Tuesday morning was hotter than ever; the mercury would hit 104 this day. In another 24 hours spot fires were spreading over 18 square miles, some six miles east and west, and three miles north and south. By now the logging towns of northwestern Oregon were in turmoil, with men and equipment pouring toward the spreading flames. Some 500 men were on the lines, still primarily in Crossett Western holdings. In the next 10 days, hundreds of fire fighters were thrown into the battle—loggers, sawmill workers, CCC boys, foresters—trucked in steady streams to the lines or camps in the vicinity of Wilark, Mist, and Hamlet. The Army set up a supply depot to dispatch food and equipment to the men. But the weather remained unco-operative; there was only scant hope of any rainfall or soothing coastal fog, and the flames were impossible to trail as they ate through the great thick stands of virgin timber, spreading now over forty thousand acres.

For many years, Mary L. Roberts had worked as "society editor" and reporter in the cramped, aging quarters of the *Washington County News-Times,* the country weekly at Forest Grove. From her rattling old upright typewriter,

---

* Since hearing Mr. Lyda's account of how the Tillamook fire started, this author has pursued the matter in detail in hopes of righting the record. From evidence gathered—written, and in tape-recorded interviews with key people who were there—I am convinced that Mr. Lyda's story is a truthful one, and that there was indeed a "second fire." However, its origin remains undetermined.

with its faded ribbon, which she often forgot to change, Mrs. Roberts wrote about the comings and goings of her neighbors, the births, marriages, and deaths to fill the columns of the weekly. Each week also, she selected the better stories to file as a string correspondent for the metropolitan *Oregon Journal* of Portland, which brought a few extra dollars to supplement her modest salary to support a family of four, now reaching college age. Mrs. Roberts was energetic, fearless, and outspoken; many times she argued heatedly with the young editor, Hugh McGilvra, who had purchased the newspaper, for she was an ardent Democrat, he a strong Republican. He could have fired her, but he held a respect for her ability as a newspaper reporter. Now in the summer of 1933, this small woman who never drove a car broke the biggest story of her career and stayed on top of it until the last embers died. Viewing the great holocaust from every angle, even plunging by car right into the flaming dragon's mouth and circling the great inferno in an airplane, which itself was still a daring method of transportation in 1933, she told about it in a descriptive, graphic manner that brought it home personally to thousands of concerned readers, developing her own unique style of coverage for that time of terror.

"The fire has been feeding largely on beauty," Mrs. Roberts wrote with feeling. "Devastation of the flames in the huge district was, and is, becoming so complete that the toll of wildlife must be immense. The dense forests provided shelter and sustenance for hundreds of deer, rabbits, chipmunks, pine squirrels, gray squirrels, grouse, partridge, and rodents, in addition to many bear and cougar. Some may have escaped by fleeing toward the outer woods, but inevitably the majority must have been trapped and burned. Without doubt, small birds by the thousands perished."

For a brief time around August 20, cooler weather allowed fire fighters seemingly to gain control of the rampage. Fifty relief fighters were dispatched from Forest Grove to allow those on the lines, including nearly 400 CCC men in 27 camps, to get some sleep. The boys, many of whom had never confronted a forest fire before, had bent their backs at great risk attempting to corral the flames. Army officers were sent from Vancouver Barracks to organize the "troops," and it proved an interesting experiment in military discipline vs. the independent civilian.

"With the respite," wrote Mrs. Roberts, "has come realization by the authorities that the CCC from Wilark, Mist, and Hamlet, undisciplined from a military standpoint, without exception revealed discipline of the finest type and a morale which left nothing to be desired. These CCC youths who made good are nearly all from the state of Missouri. They were trained at Jefferson Barracks, St. Louis. Although there is no penalty for desertion from the CCC, no saluting and standing at attention, no goose-stepping in drill, the battle of the flames was fought without confusion. Cecil Kyle, fire warden who personally led the Wilson River fight, and a deputy, Frank McCullough, declared the presence of the CCC organization, with their tools, trucks, and other equipment, which permitted mass forces to be moved quickly to dangerous fronts, undoubtedly made stopping of the fires possible as soon as the weather permitted any human agency to stop them.

"The CCC were working in groups from three to seven, each under an experienced woodsman. There was not a single accident, and only one man fell out of ranks and returned to camp in three days. He was stricken with an acute stomach ache. He rode back to camp on a truck. The spirit of the CCC boys was shown by the kidding he received as he left the fire lines. He got no sympathy.

" 'You just can't take it,' they yelled after him.

"Army officers were reported deeply interested in this research in discipline of masses of men," the newswoman continued. "Assigned to the camps without their usual authority, depending on tact and their knowledge of human nature, the officers had assumed their new duties with some misgivings. As all Army training is pointed at turning out disciplined men who will obey orders, the result has given them cause for study. The American spirit of fair play of the football field was substituted for the rigid Army practice, which has prevailed for centuries, and proved at least as effective."

Yet the hope that it was over was short-lived, for soon the flames were heating up again. Mrs. Roberts and her Portland editor made a daring trip along the narrow, corkscrew Wilson River road into the heart of the fire area, with flames leaping about them on both sides, and falling trees and limbs threatening to trap the car.

"Some of the finest virgin yellow fir in Oregon was attacked by the flames yesterday," Mary L. reported. "A bridge burned out near Shorb's clearing, effectively blocking the road to cars. Visibility was so poor that no one could tell exactly what damage had been done to one of the finest stands of timber in Oregon, or to the popular recreational district from a standpoint of the vacationist. The canyon was dense with smoke. Small detachments of begrimed, red-eyed men, laden with axes and shovels, moved purposefully through the gloom. Towering snags glowed red, showering sparks, which fell harmlessly, with no wind to carry them. Reeher's Retreat escaped the flames. Max Reeher, owner, and Jerry Weaver remained at the camp, although cut off from communication outside. Fifty men this morning were sent to Fobe Creek to throw trails around spot fires. A pack train was dispatched to the foot of Saddle Mountain, which is somewhat south of the main fire area, to trail spot fires there, which a shift in the weather might make dangerous.

"The CCC have had a real baptism of fire. During the two days and nights high winds prevailed, bursts of crown fires made them appear as insignificant as pigmies in a cyclone. Today the pigmies are winning, the cyclone having given away to a lull broken only by a crackling of burning stumps, pitch knots, and half-rotten snags."

The fire had been held to 40,000 acres when near dawn on August 24 the humidity skidded to 26 per cent and that fearsome east wind struck again, stirring the flames to renewed life. Sparks traveled afar into unharmed timber, which ignited with dynamite force and crowned again and again, leaping from tree to tree. Fire wardens, sensing new danger, ordered the full evacuation of ranchers, settlers, logging camp inhabitants, and even fire crews from the threatened region, now sprawled over several counties. They fled none too soon.

Suddenly the fire blew up—I mean that literally—on a day that old-timers still recall vividly, since few realized how awesome a forest fire can become. It exploded with the unbelievable ferocity of an atomic bomb, unknown to the world for at least another decade, and the resulting smoke cloud resembled that of an H-bomb. In 20 incredible hours the fire rampaged across 220,000 acres of the nation's best forest land, burning trees at a rate of 600,000 board feet an hour. Along a horrifying 15-mile front the fire became an awesome wall of orange flame, exploding again and again in the towering tops of 400-year-old firs, creating an inferno unlike anything since Northwest volcanic peaks erupted. The mad flames leaped thousands of feet into the air as from a billion blast furnaces, and the conflagration built upon itself, creating its own violent updrafts, which roared like a thousand hurricanes.

Three thousand fire fighters were helpless as the awesome smoke cloud mushroomed 40,000 feet into the atmosphere, visible for hundreds of miles at sea and far inland, and hanging for many days above the region. Darkness came at noon for towns surrounding the cauldron. Chickens went to roost at midday. Ashes rained down on the small towns and upon Portland. Black debris piled two feet deep on the beautiful sandy beaches of the northern Oregon coast, and fell upon ships 500 miles at sea. And adding to this horror, some incendiary with all the magnificence of human stupidity touched off another 300-acre blaze 10 miles north of the big fire, soon well beyond the control of 200 fire fighters.

"Flames could be seen last night as they leaped high above the mountains," wrote Mrs. Roberts. "The Scoggins Valley fire was fanned into new life. Innumerable small fires attacked green timber within the burned area which had escaped destruction other days. Fire in the Lee Creek holdings of the Flora Logging Company was so hot and dangerous that men were recalled to Wilson River. The weather has been playing with the fire-fighting crews as a cat plays with a mouse—let it think it has escaped, catch it—and so on ad infinitum. The present crisis is the third since the fire started. Twice the humidity crept higher, fire crews trailed and held, and then came the east wind to blow brands scornfully over the narrow tortuous ground trails to set new fires ahead. . . . Last night several miles of the nearest flames could be seen. It required little imagination to picture a fiery dragon consuming the countryside as the fire and smoke assumed the form of some dread prehistoric monster."

The first—and only—death occurred, and there were mounting injuries as the fire went wild. The lifeless body of Frank Palmer, a CCC youth from Marseilles, Illinois, was brought out to Forest Grove for shipment home. Jack Miller of Aloha, Oregon, was injured at the same time.

"Palmer, Miller, and two others . . . were patrolling a fire line two miles west of the Shorb camp," reported Mary L. "Palmer was sitting, the others standing. The tree, a large green one, was seen as it started to fall. All ran, not observing Palmer's danger. He was caught before he could get out of the way. When his comrades looked back, all they could see was one hand and his head. They dug him out with their fire-fighting equipment, found him dead, and then rendered first aid to Miller."

Thousands were on the fire lines, scattered throughout 300 square miles of wild and rugged country, and increasing the dangers of more deaths and injuries, especially with so many novices involved. They didn't understand how speedily the flames could travel in dry, pitchy timber. It worried state forester Lynn F. "Cronie" Cronnemiller, who for 10 days didn't sleep, keeping pace with the rapidly changing picture.

Two college students, James Bushong of Forest Grove and Herbert Redtzke of Portland, were trapped when a crown fire leaped a trail they had just completed in isolated country. Fire balls were bursting overhead, like flaming missiles.

"We saw a cougar and five deer chasing along together," Bushong told Mrs. Roberts. "They did not know where they were going any more than we did. We jumped into the creek. There were two deer in the creek near us. We didn't pay much attention to each other. We remained in the water until the air became too stifling for us, when we dashed for a place that had been burned over. We sat there and roasted until things cooled enough for us to get out. We don't know what became of the deer. . . ."

Again and again the fire exploded, building on its own updrafts. One woman, viewing it from a safe distance, years later remembered it as a "beautiful sight," although few would concur, in the face of tragedy. Mrs. Roberts meanwhile took to the air by private plane to gain a new perspective.

"Looking down on the week-old burn in the Glenwood area brought a sensation of dismay," Mary L. wrote. "Huge areas of ripe timber had been swept clean by flames, leaving bare ground and huge black poles, which reached gauntly upward. . . . The Scoggins Valley fire smoke presented a compelling spectacle from all parts of the valley. The east wind cleared the skies and a huge bank of smoke, resembling at times another Mount Hood to the west, and again huge, wraith-like mushrooms, was banked against a sharp horizon. Scoggins Valley timber is in rough country, cut up by canyons and ravines. Little of the flames can be seen from the valley. But from the airplane, long fingers of fire could be seen on the Wilson River eating against the wind to the northeast. These soon consolidated. The Wilson River fire yesterday fed on a three-year-old burn to the southwest, cleaning it of small stuff. Fire spread from this into surrounding virgin timber. . . . The ambitious Stimson operations in Scoggins Valley face a threat of virtual extermination. Fire is eating up a stand of timber which would have lasted the mill for a generation. . . ."

The horseshoe-shaped fire front now stretched for almost 100 miles through the Coast Range and bore down on towns and villages. Camp McGregor, one of the largest, was under siege; women and children of some 30 married men were evacuated on railroad flatcars. Some men stayed behind to fight the flames, which threatened the many buildings, including bunkhouses to sleep 200 loggers. It was in vain, for a burned tree crashed across the water line and within an hour the camp was gone. More people were evacuated by rail from the villages of Wakefield, Maples, Enright, and Mayo, the town of Tillamook was threatened, and Jewell was partly burned, while Elsie was completely surrounded, and Cochran and Timber were isolated by the flames. And fire came within a mile of Forest Grove as brands landed in a stubble field, setting blaze to the farmhouse and equipment. Still the flames swept on, with visibility down to 200 feet. Wild animals sought refuge on the outskirts of Portland, and now the railroad was out, with trestles and tunnels consumed.

After two weeks of nightmare and terror, the welcome fog and rains arrived. The fire died down, losing its punch, although the flames would continue to smolder and threaten for many months. Mary Roberts summarized it:

"With the change of wind, bringing relief of moist-laden air, lowering skies, and occasional rains, Forest Grove and vicinity lifted thankful eyes to the hills. Monday morning a changed world greeted early risers, who remembered dreams during the night past of cooling showers, which washed away a smoke-filled world, sinister and apprehensive.

"Since Friday and Saturday of last week, the fire situation has completely changed. Reports then were harrowing details of ugly flames licking greedily at tall timber, mad battles to keep fire trails open, burning areas isolated. Dogged, dreary labor with shovel, ax, and hazel hook, with weary backs, blistered hands, bitterly sore feet. And to no avail. Again and again the flames swept across the safeguards, eating, ever eating at Oregon's giants.

"With the closing down of saw and lumber mills in the state last week, the hazards of added fuel to huge fires dropped out. But with the change of atmospheric conditions, forest fire fighters and citizens of the state breathed easier."

The devastation was indescribable in the millions of black and ruined trees; charred, grotesque carcasses of thousands of animals; and the eerie desolation of the entire region, an area about half the size of Rhode Island, totaling 311,000 acres containing some 12,200,000,000 board feet of timber—enough to supply all the sawmills of the nation for a year and sufficient to keep all Portland sawmills operating for two decades. And most of it—220,000 acres—was consumed during that single big blow-up of August 24–25. The full impact of that gigantic loss in timber value, $200,000,000 in payroll, and the creation of a worthless region staggered the imagination. The Green Desert was truly a hopeless desert now, and there were many who felt that it would never come back. It would take time for the horror of the disaster to be realized in total, for it would substantially alter the complexion and economy of the state and region, and if you can believe old-timers, even local weather patterns.

Still, the timber lands continued burning at an alarming rate, despite the shock of the Tillamook Fire. Three years plus a month after that milestone, fire again went on the rampage along the southern Oregon coast, this time destroying an entire town, trapping its inhabitants on the beach, and costing many lives. Again, logging operators were blamed.

As was the usual practice, loggers set slashing fires to clean out the debris when fall approached and the weather became reasonably "safe." The pungent smell of wood smoke had become an accepted part of that time of year, the true fragrance of a Northwest autumn. But in 1936 the September days grew unseasonably hot and dry. Once slashing fires were set, they were difficult to extinguish, sometimes even to contain. This was what was happening near the beach resort of Bandon in the latter part of the month.

On the morning of September 26, Bandon was struck by a series of isolated but unexplained grass fires. Some people wondered if youngsters were having a spree with a box of matches. But conditions were dynamite: The humidity was down to 8 per cent and the mercury was climbing, at the moment hitting 80 degrees, which is hot for the Oregon coast.

Several slashing fires were breaking their bounds, according to reports, while a heavy smoke pall hung ominously over the region southeast of town. There were more small grass fires as the day wore on. Bandon fire chief C. S. "Curly" Woomer suspected their origin was sparks from those slash fires. The town was particularly vulnerable, too, for throughout this coastal resort, in nearly every vacant lot and surrounding the homes and cottages, grew a low-slung prickly bush called Irish gorse. It was imported by the founder of Bandon, Lord George Bennett, to beautify this Oregon village in a manner reminiscent of his native hamlet in Ireland. And in springtime the yellow blooms of the gorse made it a dazzling spectacle. However, Bennett sowed the seeds of his town's destruction, for the gorse was highly flammable in hot weather from the oily substance dripped by the leaves.

The uncontained slash fires crept closer to Bandon and the spot fires became increasingly serious for Woomer and his volunteer firemen. By early evening, the gorse was catching fire. Suddenly it exploded along a residential street in the southeast part of town. Volunteer firemen tried to contain it there, but sparks were rapidly igniting fires all over that end of town, while residents, unaware of the sudden danger, crowded into the local theater to see a new film entitled *Thirty-Six Hours to Live.*

The show was halted by the chief, calling for volunteers. But nothing could stop the fire now. It roared through the town's upper level, destroying every building but the high school. As it reached the bluff and bore down upon the city center, people were fleeing for their lives. Several had already burned to death. Their only chance was to reach the beach, since they were trapped on three sides by the fire closing in on them pincers-like from the south and east, with the deep Coquille River on the north. In cars and on foot, they streamed toward the ocean as fire balls of flaming gorse fell about them and set the driftwood ablaze. On the beach some 15 cars were afire, their gas tanks exploding, creating additional danger. Frightened men, women, and children stood waist deep in the surf, holding hands, and watched the town burn in fury, while flying sparks set the grass ablaze on offshore rocks behind them.

The flames roared on throughout that terrifying night. When nothing remained to burn, the fire died out. Bandon resembled a bombed and napalmed Southeast Asian village. Only the grotesque skeletons of its few brick and mortar buildings remained. Twenty-three persons were dead, while 400 dwellings, eight churches, a grade school, and all but one business building were leveled. Almost everything was reduced to smoldering ashes; only charred chimney spires marked the graves of once-fine and happy homes. And Bandon, which managed to rebuild, wasn't likely soon to forget.

The pressures were mounting to do something about forest fires, both from visionary leaders within the timber industry and from the public, which cursed the recklessness of "those goddamned logging outfits." Public animosity led to stringent new laws and regulations in state legislatures; some of the new acts of legislation were unacceptable to the industry. The problem was that it was such a huge and varied logging show, involving big companies and little individual operators. Then there were the railroads, the hunters, the fishermen, the hikers, the motorists, and all those who thoughtlessly left campfires burning and tossed cigarette butts out car windows or into tinder-dry brush off hiking trails. What could be done about them? And where do you begin?

Yet the Tillamook Burn served as a haunting reminder to all who drove or hiked through that lifeless region, with its thousands of ghostly black candles, now gradually becoming bleached a gray-white as they marched up the rough slopes for hundreds of square miles, as far as the eye could see. And every six years, with weird and ironic consistency, fire exploded again in The Burn, blacking additional green acres and acres trying to come back, and destroying what few trees managed to survive. The Burn was cursed by a six-year jinx, loggers said, since it happened four times into the 1950s. The doubt was very strong that The Burn would ever be a forest again.

Nevertheless, a lot more than trees and dollars and a town went up in the flames of 1933 and 1936. The old ways went with them; the seeds for a new style of thinking, an entirely new approach to the forests, were sown amid the bone-white snags and skeleton remains of those milestone conflagrations.

PART TWO

NO CORKS ALLOWED

Chapter X

Big Park and New Idea

Why "tree farm"? The moment I say "farm," there immediately flashes in the mind of every person here a mental picture of his own personal idea of a farm. A farm is an area where a man grows successive crops of corn, wheat, hay, oats, and vegetables. We co-operate with farmers because they harvest the crops that feed the nation. Well, timber is a crop, just like any other crop, except that it takes a longer time to grow a crop of trees. So why not talk about and have tree farms?

RODERIC OLZENDAM
Dedication of first tree farm
Montesano, Washington, 1941

Since 1897 a mighty mist and rain-drenched section of the green velvet land of the Olympic Mountains, one of the West's most formidable ranges, has been under the protective wing of the federal government, first as a preserve, then as a monument, and finally as a national park. About two thirds of the Olympic Peninsula bounded by Puget Sound, the Straits of Juan de Fuca, and the Pacific Ocean—some 2,188,800 acres—came into U.S. custody in February of that year when President Grover Cleveland signed a proclamation placing 2,883 square miles and 61,000,000,000 board feet of timber under the first true controls over the nation's remaining timber supply. Another 20,000,000,000 board feet stood outside the preserve, much of it within sound of the mounting number of sawmills at Grays Harbor. The Olympic Reserve was among several proclaimed by the outgoing President as one of his last legislative acts.

The action held significance, since it appeared that the warnings of naturalist John Muir, like those of Rachel Carson of a later generation, finally were sinking home to the American people. Muir was simply saying, again and again in straightforward terms, that we would one day run out of timber at the rate it was being cut, and that the ramifications from such a state of affairs would be felt by many generations of Americans. Yet for a very long while, Muir was shouted down; even if it were true, there was no way of turning things around in an age of exploitation and the gutting of the nation's natural resources—in minerals, soil, and forests.

Now the answer seemed to be, "Lock it up." The lumbermen cried "foul" and much stronger epithets, and heavy pressure was put upon Congress, which suspended the Cleveland Proclamation for nine months, giving gleeful speculators and land grabbers an opportunity to make off with at least a portion of this lush, beautiful forest realm, unlike any in the world. President William McKinley reduced the reserve's size by 750,000 acres. Railroad land companies swapped sagebrush and wasteland for equal acres of the Pacific Northwest's green desert, which they set down as a "fair exchange." The jockeying, manipulation, and exploitation over the grand peninsula's natural wealth was something to behold during the next few years, and finally emerged the first suggestion that the region be preserved for all time as a national park.

Certainly of as much concern as the timber and the beauty of the region was the preservation of the herds of Olympic elk that roamed the primeval forests, yellow meadows, and alpine slopes in the shadows of towering 8,000-foot Mount Olympus, that formidable peak noticed in 1774 by a wandering Spanish sea captain, Juan Perez, against a dazzling sky of brilliant blue. The elk herds were being butchered like the buffalo of the plains by professional hunters to sell their teeth as watchfobs to members of the Benevolent and Protective Order of Elks. Benevolent? Protective? The Washington state legislature in 1905 tried halting the slaughter by outlawing elk hunting, but its action did little good. The poacher shooting and dental extractions went right ahead. Efforts were made to establish an "Elk National Park" and also a game refuge, but these were also struck down. Then William Humphrey, a congressman from Seattle, pounded on the White House door of President Theodore Roosevelt, obtained his ear, and told the story of the doomed animals. Roosevelt was the right man for the moment, since his own concern for the outdoors and the American environment had created a sweeping conservation movement. Like Cleveland, Teddy was about to leave office, but he moved quickly, writing an executive order establishing Mount Olympus National Monument, which embraced 620,000 acres, half in forest, from which hunters, loggers, and miners were barred. The outcry was very shrill, as might be imagined, but the elk were saved. Then, under the pressure and frantic patriotism of World War I, the locked-up territory was halved to 328,000 acres, mostly above the timberline, thus making rich forested areas of virgin growth once again ripe for the plucking.

The Grays Harbor logging outfits were creeping up the peninsula, pressing hard around the boundaries of this locked-up timber supply, and nibbling at it wherever possible. The years of the Big Cut felled giants by the thousands—and they were whoppers. An early photograph on the Satsop shows 28 people standing on a single stump. Spruce log rounds were turned into tables and chairs by the settlers, while the chips of a single giant, sawed with misery whip 20 feet above ground to avoid the swollen, twisted base and the main pitch stream, created wood waste to fill a wagon. The logged-over land appeared valueless, since it wouldn't grow crops save for small truck gardens of the scattered and ambitious settlers who labored long to burn, dynamite, and grub the stumps and boulders. Most of this new wasteland would remain idle for several generations before the timber reached maturity again. And all the while the Grays Harbor mills, tooled to handle the big stuff, were as hungry as ever.

There was a movement again in the thirties to create a national park by expanding the national monument lands set aside by Teddy Roosevelt. The proposed 1,000 square miles would save most of the big virgin trees and retain things in their natural condition, for while a federal forest may be submitted to logging, a national park allows nature only to take its course, presumably forever. Park forests are permitted to go through their entire life cycle, to mature and die, then rot away as a new forest springs up without the helping hand of man. The timbermen call it waste; preservationists maintain it isn't waste to retain some areas of the continent as they have always been for future generations to view and enjoy, and be reminded of what the New World was like before the destructive European landed. But the national park scheme didn't always work to advantage, for Nature herself was often cruel and wasteful, as she had been on the Olympic Peninsula in 1921, when 110-mile-per-hour winds slashed a swath through the forests 70 miles long and 30 miles wide, destroying some 6,000,000,-000 board feet of timber. That windstorm also killed thousands of elk and set back timber harvesting plans for a generation.

The proposed Olympic National Park would lock up, nevertheless, some 13,-000,000,000 board feet of top-grade timber. When legislation was first placed before Congress in 1935, it touched off a bitter and rowdy feud, which extended over the next several years. It was spiced furthermore by a heated internal ruckus between Secretary of the Interior Harold Ickes and Secretary of Agriculture Henry Wallace during the administration of Franklin D. Roosevelt. Administration of national monuments had been shifted from the U. S. Forest Service to the Park Service, which meant that Ickes the conservationist gained control over all those wonderful trees. And as he charged bitterly at the hearings, "The ultimate purpose of the National Forest Service is to grow and develop trees so they may be available for use in the lumber industry." Furthermore, he alleged that "if the exploiters are permitted to have their way with the Olympic Peninsula, all that will be left will be the outraged squeal of future generations over the loss of another national treasure."

The pro and con arguments over the park were the same ones heard time and again across the decades, into the 1970s, when another sweeping conservation movement would attempt to preserve from cutting other great last stands of wilderness. It is the reason many people loathe history, reject it, and try to bury it, for the echoes of the past rise to haunt each generation of men by showing them that we haven't progressed as far as we like to think, that ideas and viewpoints are not necessarily new, but only another round in what is an ever-widening circle. The proposed park was too big, it was argued at the congressional hearings, and the rainfall far too heavy for the enjoyment of tourists. The trees were now overripe and needed to be cut; to do otherwise would be pure waste of the nation's timber supply and its vitality. The counties needed the tax money gained from the stumpage fees. The woods wouldn't be wiped out if logged under Forest Service regulations. But park advocates, seeing those trees from another angle, retorted, "To a lumberman a forest is preserved when trees 600 years old are cut down and followed by pulpwood saplings cropped every 40 years."

Moreover, there was the serious crisis at Grays Harbor: The timber was truly running out. The many sawmills, which were geared for big logs, were hard-pressed to keep operating, and now the do-gooders and conservationists wanted to padlock all those great trees to the north. If that happened, the mills would be shut down and hundreds of men would be thrown out of jobs. The Harbor would become a chain of ghost towns. But park supporters argued that at the rate Grays Harbor gobbled up the timber, what remained would last at best only another decade. Then the mills would be closed anyway and the workers forced to move elsewhere. Was it worth it, they asked, to destroy this primeval wilderness of beauty and spectacle, showing Nature in her finest hour, for a few more years of cutting?

The House committee thought not, and approved the legislation. But Congress didn't act, for there was more maneuvering, politicking and horse-trading with the powerful timber lobbyists. President Roosevelt toured the Olympic Peninsula in 1937, the idea of the timbermen being that he might change his mind about need for the park. But the tour did the opposite, especially after viewing the sad ugliness of slashed and logged-over lands near Grays Harbor.

"I hope the son-of-a-bitch who logged that is roasting in hell!" Roosevelt growled within earshot of reporters.

The remark, well publicized in the press, set the timber industry back on its calked boots. Roosevelt's attitude was hard to the core. The following year Mon C. Wallgren, a congressman from Everett, the big mill town, submitted a new bill for a 910,000-acre park amid the protests of Governor Clarence Martin. Harold Ickes advised that it would be better to work for a smaller park, containing a stipulation that the President could expand it through proclamation. The Wallgren Bill passed; Roosevelt signing it at the end of June 1938, creating a park of 643,000 acres, which could be expanded to 892,000 acres. Within two years Roosevelt added 187,411 acres, boosting the total to 830,411 acres or 1,308 square miles. That expansion touched off another round of controversy over the billions of board feet of evergreen; once more the battle raged over a vanishing wilderness, containing two-mile strips of the beautiful Queets Valley, aimed at keeping one river in its entirety and also a coastal frontage section for the public. The idea of preserving scenic rivers and coastal beaches isn't therefore something new for the 1970s, but has been around awhile. Finally, President Harry Truman arranged for federal purchase of the Queets corridor and the coastal strip through presidential proclamation, boosting the park's acreage almost to its allowable limit of 892,000 acres.

Since then, efforts have been advanced from time to time to unlock the great park for its timber, to reduce its over-all size, or to thrust paved highways through its heartland. Arguments are that it is much too large and its access limited only to those able to negotiate rugged back-country hiking trails rather than enjoyment by the masses in their polluting automobiles. Again and again its protectors have arisen to shout down those desiring to open up the big park, among them Supreme Court Justice William O. Douglas of Washington State, who has often hiked its back trails and written fluently about its grandeur. So far it has remained intact, one of the greatest remaining wilderness areas on this continent.

Nevertheless, Olympic National Park and the adjacent national forest brought the lumbermen solidly against some cold realities. They had their backs to the sea, cutting the last big stands in Washington State, and many now shifting operations into Oregon (later it would be British Columbia and Alaska), where other lush forests stood tall in the Coast Range Mountains. Unhappily the cutting, the waste, and the fires would go on there as they had in the past. However, the lumbermen sensed a marked shift in public attitude toward the nation's resources. The battle for Olympic Park and its final establishment awakened the timbermen to something a few out-of-step leaders had harped upon for years. The park was symbolic of a new age; the American people were becoming increasingly concerned over what was happening to their resources. The wilderness was being pushed back; the forests were the most visible and accessible, and therefore a natural target for any conservation movement. Yet certainly the timbermen themselves weren't wholly unaware of the problems. Scattered through the Big Woods were thinking men who realized that logged-over lands shouldn't be abandoned, but somehow put back to work, and that someday soon it would from necessity have to be done on a massive scale.

Before World War I, up in the Grays Harbor country, Alex and Robert Polson were experimenting with pioneering methods of dealing with logging slash so as to encourage new growth. When fire blackened an area where there were small trees, the Polsons sent out their own loggers, paying the wages, to collect cones and scatter seed on 2,600 charred acres to help the land come back. Sol Simpson was also saving his lands, for reasons few in that era could understand. In the deep woods tough loggers and lumbermen with rough, callused hands gently set out tiny tender seedlings as early as 1905. Sometimes they were kidded by fellow lumberjacks for doing "sissy's work," and also, why bother? But offbeat lumberjacks agreed with the "bug men," as they called the early trained foresters, that a beginning was necessary to replenish the forests. Therefore, "tree farms" were growing here and yonder long before such a term or such a movement was introduced.

Despite public and political outcries that they were evildoers, many loggers and lumbermen held a deep love for the forests and a realization of their continued importance as a basic national resource. One of the key problems which discouraged timber operators from retaining their holdings, of course, was having to pay heavy taxes on logged-over land. It was simpler just to let it go, for the investment on the long haul was too much to carry. If lumbermen were to be encouraged to replant their lands, significant changes were needed in the tax laws.

George Long, Everett Griggs, Mark Reed, and other timber leaders had long recognized that time was running out, but if the conditions were right, it would be possible to bring back the forests of this mist-washed Northwest country within a comparatively few years. However, until the woods were made reasonably safe from fire, any effort toward costly reforestation was a ridiculous pie-in-the-sky. State laws were lax on fire safety, and, contrary to the popular view, fires weren't by any means always the fault of the operators or reckless loggers, for it stood to reason that they would wish to protect their investment and the very thing that was giving them a livelihood. However, regulations were almost nil, even to requiring spark arresters and other safety devices on motorized equipment, including railroad locomotives. But the public must also shoulder much of the blame. Summary reports showed that half the forest fires annually were caused by hunters, anglers, campers, hikers, and other segments of the population. Yet anytime a fire broke, it was automatically blamed upon the loggers, and the notion was so strongly imbedded in the public mind that it would be difficult to shift indifference to responsibility.

"Forest protection means that reforestry is practical, and especially on the Pacific Coast, where nature has already grown the greatest forest known," George Long told the Pacific Logging Congress in 1911. "Nature will do her part again if man will do his. Lumbering can be made perpetual on the Pacific Coast, and we can transmit to our sons and their sons an opportunity to follow our chosen profession. The logger and the lumberman will be the true conservationist of the future. . . ."

Two years before that speech, Long founded the Western Forestry and Conservation Association, which became something of a clearinghouse on matters of forest protection and practical timber growing.

"Help lumbermen with forest fires and forest taxes, and they will find a way to regrow the timber," he told U.S. congressmen in a plea for "understanding." "We are anxious to get into this reforestation game. We realize the necessity for it very keenly. And out here where the West ends, we want to begin to grow a new forest and will do it when we have the slightest chance of making it a profitable enterprise."

"Profit" was of course a dirty word with the lumbermen's critics, but what Long advocated made good sense, since that was the way things were accomplished in the United States. Major Griggs concurred; the Tacoma lumber king had never accepted the cut-out-and-get-out philosophy, but felt keenly a responsibility for the land and its future. He wouldn't default his land for taxes, since Griggs viewed the manufacture of lumber and the growing of timber as a single enterprise, as with a farmer in harvesting and replanting. He asked the Western Forestry and Conservation Association to survey and classify his holdings, and then hired one of its men as a "company forester," in a strange departure from the old-style timber cutters, who sneered at the college-educated bug men.

Griggs' enthusiasm grew as he tramped through the woods to learn about reforestation and growth rates. He instilled this same enthusiasm in his crews at St. Paul & Tacoma, and hired high school boys and back-country homesteaders to plant seedlings. Signs were posted along back roads of his holdings, "New Forest Growing Here," and one tract was set aside for schoolchildren through the efforts of Mrs. Griggs, for she agreed with her husband that young folks needed to be taught a healthy respect for the timber lands. By the time Major Griggs retired in 1933, 70,000 acres, constituting the company's original holdings, were blocked out into one of the most productive replanted forests in the Pacific Northwest.

Then there was Mark Reed, a native of Olympia and foreman of the mighty Simpson Logging Company of old Sol Simpson, who came to Mason County on the southern nudge of Puget Sound in the 1880s to help build a logging railroad, then turned lumberman. Simpson was an innovator, using horses instead of bulls and thereby getting out the logs faster. He went against the grain of accepted tradition in other ways, but his strangest practice was hanging onto his lands, while others sold them to stump ranchers or let them go.

As in other instances within the lumber kingdoms, notably Pope & Talbot, Mark Reed married into the company and rose to general manager, then president of the big show at Shelton, Washington. He shared the belief of his father-in-law that the land might one day be productive again, as it was indeed doing on its own, slowly but surely, with no special effort on the part of the owners. Little trees were sprouting up, for the powers of the Northwest forests to come back are amazing things. Reed also took special interest in developing the rough lumber town of Shelton into a fine industrial center. He and his associates presented the city with such basic necessities as a high school, library, hospital, and memorial building. Reed was a progressive; the company had a responsibility here, beyond merely mining timber. But what would become of Shelton once the timber were gone? The pattern was clear to Mark Reed: Back of the community was the busy industry, and the prime need behind that industry was a perpetual forest. None of the three could long survive without the others.

Reed knew what must be done, so he hired a forestry-schooled engineer of the U. S. Forest Service to work out a permanent program. Additional lands were purchased until Simpson's had 150,000 acres for growing trees. But this wouldn't work if other logging ills weren't corrected, so Reed teamed with George Long in writing Washington's early lumber code. Then he served 17 years in the state legislature, helping to put into law sensible conservation practices which he'd learned in the Big Woods.

But the man who pulled everything together into a new age for the forests was Colonel William B. Greeley, one of Gifford Pinchot's Boys, who broke with the famed first chief forester over philosophy and became a controversial figure with conservationists because of his more lenient attitude toward the timber industry. Greeley later became U.S. chief forester, but that didn't alter his partnership approach toward the industry. While Pinchot saw lumbering as a "willful industry," Greeley viewed it as a "sick industry" that could be cured.

Fire was Public Enemy No. 1 against growing forests for the future, and Greeley knew all about fire. Early in his career in Montana, he'd been head of a newly created district of the Forest Service when 3,000,000 fine acres were destroyed and 85 lives lost. From that moment he realized what a tremendous task faced any movement toward sensible forestry. And with the clairvoyance of youth, he swore to devote his life to driving fire from the woods—something that in 1910 seemed utterly impossible, for in most circles, fire was considered among the hazards of living, like today's traffic accidents.

There was something else: Greeley believed that a common meeting ground was needed for the enemy camps of lumbermen, foresters, and the public. They all had that mutual foe—fire. It could bring them together. This enemy, Greeley thought, couldn't be defeated by any single group or through hostility, but only through the co-operation and combined efforts of the public and the industry. He cinched it a notch farther: if this could be accomplished, it might just create an area of understanding and bonds of partnership that would take forestry another step toward the ultimate goal—reforestation.

Greeley and Pinchot broke over this very matter. Pinchot stood firm that the lumbermen would never mend their unruly ways and that so far as talk of reforestation was concerned, the timber barons simply weren't interested in any long-range investment in the care and protection of little trees that wouldn't be worth anything in their lifetimes. The odds were heavily against those trees ever reaching maturity anyway, and if they did, well, that was a good many years away. But Greeley worked hard, often gaining the strange co-operation of state foresters and lumbermen for tempered legislation, rather than the Big Stick tactics advocated by Pinchot, who was now out of public service but still a very powerful voice among conservationists. What Greeley feared were laws that would strangle the industry, although he concluded that some regulation over destructive logging was inevitable. While chief forester, he urged the Forest Service to work with lumbermen and state forestry departments to develop minimum requirements in cutting practices.

"My hardest work," Bill recalled later, "was to keep the lumbermen in the co-operative tent. They might easily have stampeded off the reservation and taken to the warpath. Some of them did."

Greeley found a sympathetic ear in a newcomer to the nation's capital, Senator Charles L. McNary of Oregon. McNary asked what could be done. Greeley expressed his fears of federal police action in the woods, which would solve nothing and only create bitterness and rebellion among the timbermen. The crying need was for practical, sane industrial or private forestry, and this was heavily dependent upon forest fire prevention and the style of forestry taxation. Senator McNary rose to the occasion, demonstrating the kind of foresight that gave him a long and celebrated career in the U. S. Senate. Why not, he inquired, write a bill around forest fires?

The Clarke-McNary Act, ready early in 1924, was based on co-operative action for fire prevention, but it extended into forest management and tree planting. It encouraged states to adjust taxation, allowing private interests to carry forward programs of commercial reforestation, increase nursery stock where states weren't able to meet their needs, assist states in acquiring more forest lands, encourage small farm woodlots, and provide for the expansion of national forests. The act also paved the way for an industrial forestry program through a co-operative system between the two key Pacific Northwest timber states.

Four years later, the McNary-McSweeney Act established experimental stations in 14 forest regions supported by an annual federal appropriation of $1,000,-000. These twin pieces of legislation showed remarkable understanding on the federal level to the problems of the timber regions. They cleared the air by changing the picture of private and public forest interests, committing the states and the federal government, plus the lumbermen, to the reforestation of denuded lands. Despite the general lack of effort in the past, the hearings disclosed that an amazing number of lumbermen were already growing trees and/or holding onto their lands with the hope of reforesting them. By comparison, only 7,051 acres of national forests in 11 states had been replanted in 1922 with new trees.

"The conviction became strong in my mind," declared Greeley, "that industry itself was the greatest latent force for reforestation if the government could shake off some of its shackles."

Time and again Greeley pounded home the key to the future in four words: *"Timber is a crop."*

With this in mind, the lanky woodsman resigned in 1928 as chief forester to accept the position of secretary-manager of the West Coast Lumbermen's Association in the Pacific Northwest. His enemies winked I-told-you-so. Greeley had sold out to the industry, which proved what they'd always suspected about his leanings. But Greeley shunted aside such criticism, for he saw a far broader picture and felt he could do a better job for all concerned, notwithstanding the American people, from within the industry than always as the potential enemy in government uniform. The timber industry needed many changes, and Greeley wasn't one to idle behind a desk. He had on his office wall the motto, "The busi-

ness of life is to go forward." His immediate goals were to move forward against
fire in the woods and simultaneously to bring about widespread reforestation
through some style of national movement. In these areas he had a meeting of
the minds with that scattered group of Northwest timbermen who for years had
been harping on these very points. But much needed to be done to alter not only
the thinking of lumbermen and loggers, but also the attitude of the public and
government agents before there could be hope for a sound future in the Northwest timber world.

Yet the temper of the woods was changing significantly. Confronted by continuing disasters like the great Tillamook Fire, mounting public animosity, and
rapidly diminishing timber supplies, the lumbermen were beginning to make an
about-face. There was a serious stirring within the industry, from which any
drastic reforms must finally come, for in the Northwest, lumber was king. Almost
all of life was geared to this mighty industry, where even the highway traffic laws
and hauling regulations were tuned to the roar of the great logging trucks, which
were becoming more prevalent with each passing year and were knocking out
the railroads. But the timber people were being moved along by the philosophy
and ideas of Bill Greeley, working through his active and talented staff at WCLA.
Legislation was encouraging a positive shift in emphasis, although fire and taxes
remained the principal problems that appeared insurmountable among many
timber people. However, in 1931 Washington State passed a law allowing taxation of cutover and regrowing timber lands on a bare assessment, linking it instead to a tax yield on any future crop *at the time of harvesting*. This would
produce a whole new ball game. There was also hope that forest fires might be
reduced through much stricter laws against logging operators and increased fire
protection. However, there was that other big factor—that public minority,
which also caused many rampaging fires for which the industry got the blame.
Greeley was doing a lot of thinking about that.

Then in the summer of 1941, a few months before Pearl Harbor, the dream
that had been kicking around for half a century, often coming up for talk in the
saloons and hotel rooms, reached a historic milestone at Montesano in the Grays
Harbor country, where a sizable crowd gathered to hear Governor Arthur
B. Langlie dedicate the nation's first bona fide "tree farm." It would be called
the Clemons Tree Farm for an early lumberjack, Charles H. Clemons, who
logged much of the region. But the dream extended back to Sol Simpson and
other rugged timber barons who, long before the Greeley years, maintained that
"timber is a crop" that could be harvested and regrown again and again in this
wonderful region so ideal for producing healthy trees. Now the Weyerhaeuser
Company was committing 62,241 acres, almost half, of a total 130,000 acres, to
a gamble with the future. The remaining land was the intermingled acres contributed by other timber companies, the county, and the state. Within 10 years
this very successful farm would be more than doubled, to 327,000 acres.

This wasn't a publicity stunt, although foresters and lumbermen hoped to gain
a lot of mileage from the event. The name and the idea must be gotten across to
the public, where there was much need for understanding. Nevertheless, the
catchy identity was wisely chosen. The term "tree farm" spread quickly across
the continent, for it was something the people could readily understand. Actually, the twin words had been kicked around for years, being used by Gifford

Pinchot, who declared that "wood is a crop, forestry is tree farming"; by Colonel Greeley, who saw the title "The Tree Farmer Gets a Chance" in a magazine and filed it in his mind; and by the lumberjacks' own author, Stewart Holbrook, who referred occasionally to the logger as a tree farmer. Still, the lumbermen needed reassurance, and Chapin Collins, then editor of the *Montesano Vidette,* gets the credit for sending the term on its way when Weyerhaeuser foresters were trying to agree on an all-inclusive term.

"Call it a tree farm," Collins said confidently. "Put tree farm signs all over the place. People know what a farm is. They'll understand what you have set out to do."

The term fit, for the general idea was to appeal not only to big and small timber operators, but to farmers and others owning small woodlots or open acres where trees could be grown, and where the owner could qualify for tax adjustments. That autumn of 1941, a special committee met in Portland under the auspices of WCLA and the Joint Committee on Forest Conservation, which became the Industrial Forestry Association. The outcome was a system to certify potential tree farmers who pledged long-range forest management of their properties. The tree farm was readily defined as an area, no matter of what size, dedicated "to the continuous production of forest crops." Would-be tree farmers found that the requirements were much more extensive than merely setting out seedlings. They were told to furnish intensive protection not only against fire but also against tree diseases, excessive grazing, and insect infestation. An owner needed to satisfy the committee that his intent was strong and earnest, and his plans sound, which often required not only a personal appearance before the conservation group, but an on-the-scene inspection by a qualified forester.

There was war again in Europe and the nation was "arming for defense," but this was still a big year in forestry. In addition to launching the tree farm movement, Oregon passed the nation's first forest reseeding law, and an industrial forest nursery was established on Nisqually flats near Olympia. Both were integral parts of the tree farm idea. The reseeding law was written, significantly, by the loggers and lumbermen themselves, working closely with ranching and agricultural interests. Its basic purpose was to bring under control the industry's irresponsible minority, which clung to old-fashioned cut-out-and-get-out ideas. The state was given police power over all logging operations. Under its surveillance, at least 5 per cent of the original stand must be left for natural reseeding, and in the pine country, all trees under 16 inches in diameter. This Oregon law, along with a similar act in Washington State, would keep control over timber lands other than federal forests; and both lumbermen and state foresters were anxious to maintain home rule rather than be dictated to by federal bureaucrats, who often exhibited scant understanding of the problems.

Stewart Holbrook hailed the new law as "not just another attempted crackdown by government on the lumber industry, which often and many times has served as a whipping boy for the federal administration since 1933." This time the industry and states were out ahead, shaping the future of the forests in a workable and sensible manner. It is significant, too, that it took place in Oregon, which had now exchanged places with the Evergreen State as the nation's leading lumber producer, with an annual cut of some 4,500,000,000 board feet. The industry's core had shifted south, for big timber was a fading memory on Puget

Sound, where Henry Yesler, George Bush, the Popes and the Talbots cut the first boards, and was running mighty low in the Grays Harbor country.

That 20-acre co-operative nursery at Nisqually was largely the brainchild of Corydon Wagner of St. Paul & Tacoma, the current president of the WCLA, nephew of Major Griggs, and a graduate of the Yale Forestry School, also something rather new to the lumbermen's circle. The $200,000 nursery site was carefully selected after many test borings, which disclosed the land to be very well suited for the growing of seedlings. It would produce some 5,000,000 seedlings annually, and Wagner proudly revealed that already tree farmers were standing in line and that there were on hand contracts for 21,500,000 trees—80 per cent of the nursery stock over the next five years in Douglas fir, Port Orford cedar, western hemlock, and Sitka spruce. In a few years the nursery's size was doubled. Appropriately a farmer, Earl McDermitt, was placed in charge. In 35 years' experience, McDermitt had grown most every kind of crop. He proved his worth: Within eight years he grew some 32,000,000 seedlings for the industry, successfully stepping the rate to about 8,000,000 seedlings a year, then to 10,-000,000.

The tree farms, the progressive ideas, and the education of the public and the industry were all brought into a co-ordinated effort by Colonel Greeley through his West Coast association, which was growing in influence. Greeley surrounded himself with people of ability and talent who could tell the story of the changing forest attitude in a popular manner by their words, photographs, artwork, speeches, and radio programs, via the nimble typewriters of Stewart Holbrook and James Stevens, who became nationally known in their own right through their magazine articles, newspaper features, and books, many of which were devoted to some phase of life in the tall timber. Holbrook grew in stature into one of the nation's leading historians, especially with his books on logging and lumbering, while Stevens set down for the first time the wildly humorous tall tales of Paul Bunyan and his Blue Ox, Babe. Later, too, came Arthur W. Priaulx, a Eugene, Oregon, newspaperman who coined the word "litterbug" and who for many years served as WCLA's chief press agent and huckster, interpreting the industry's role in World War II and organizing educational tours and planting expeditions for school youngsters. The second world holocaust gave everything a setback, for the lumber needs suddenly became critical, and plans for future forests had to be shelved to get out timber for war as rapidly as possible. But at least a start had been made.

Despite the war, the tree farm movement grew quickly beyond the industry's hopes or expectations. There were 20 farms totaling 1,620,533 acres that first year. The program steadily developed until in the single peak year of 1956, 100 new farms were certified. By 1960 there were 628 certified farms spread over 5,872,564 acres in the Pacific Northwest. What was equally surprising was that the farm woodlot surpassed the timber companies in total acreage.

The movement spread across the continent to 29 states by 1950, covering 22,000,000 acres and inspiring slogans such as "Trees for America," which brought about a new public awareness about the dawn of a new forestry age. In 1925 only 1,600,000 acres were planted voluntarily by man, but by 1968 30,000,000 acres had been reforested, with Florida, Georgia, Mississippi, and

Oregon the pace-setters in some 5,000,000,000 seedling trees. Landowners in 1970 planted 1,206,700 acres in trees, Oregon leading the nation with 162,375 acres, and Florida second with 160,745 acres. By 1972 there were 31,474 tree farms of various dimensions across the conterminous United States (as yet none existed in Alaska and Hawaii) for a total acreage of 74,309,767. While the movement stayed as strong as ever in the Pacific Northwest, other latecomer timber regions more gradually turned to tree farms as the answer for future lumber and wood by-products. Some outdid the originators; the largest farm in Maine is a grandiose spread of 1,408,000 acres.

The Douglas fir region, which blazed the comeback trail, has 927 farms totaling 7,412,494 acres, ranging from under 10 acres to a whopper of 500,000 acres near Mount St. Helens in Washington. Counting everything, the two Northwest states have 1,386 tree farms today, representing an acreage of 10,524,910, including the magnificent Clemons tract, where it all started.

And the tree farms are getting healthier all the time, what with advancements in fire and disease control, and development of fast-growing seedlings. Despite rapid urban growth and rising population, Americans appear to enjoy living on top of each other, which is fortunate for forests of the future. Great portions of the nation's land remain timber green, for even New York State has 45 per cent of its area in trees, while New Hampshire leads everybody with 84 per cent. In three decades, the tree farm has spread a far distance from rowdy Grays Harbor.

Chapter XI

Use Your Ash Tray

This program was made for one man—Stewart Holbrook. . . . That was the only good sales talk I ever made in my life. I could hardly believe it when the governor said, "I'll take him." Then Holbrook had to be sold. . . .

JAMES STEVENS
Olympia, Washington, 1940

Summer after summer the fires continued unabated throughout the 1930s. More thousands of rich virgin acres of the Green Desert went up in smoke and ashes, advertising in massive billowing gray clouds that the region's timber wealth was being destroyed in an appalling fashion for which there appeared to be no realistic solution. The lesson of the Tillamook and Bandon holocausts seemed not to have mattered much, for the recklessness remained, within and without the timber industry.

In one summer alone—1938—more than 250,000 acres were blackened in Washington State, while Oregon sustained 2,500 separate fires. Unhappily, many of them were man-caused, the term applied to fires that originated from human carelessness rather than from the activities of the logging industry. There was reason for this: Great portions of the Pacific Northwest were still largely frontier wilderness, completely divorced from the urban East Coast, with sparse populations living in scattered villages and on isolated homesteads with a kind of pioneer independence and defiance of restrictive laws and regulations that has long been characteristic of this region. Nevertheless, the public fixed the blame automatically for all forest fires on logging, and particularly the maverick gyppo operator, who exhibited a swaggering damned-if-I-care attitude toward both the industry as a whole and the public. He owned his land, he would log it how and when he wanted, and he defied any and all comers to try to stop him.

Yet there were other causes. Railroads operating without spark arresters, and gasoline-powered trucks now invading the forests could touch off a disaster in low-humidity weather. Albert Wiesendanger, a forester at Eagle Creek in the Columbia Gorge, remembers how in extreme fire weather, he could go onto the Bridge of the Gods at Cascade Locks, where he held a commanding view along that deep defile. From this vantage point he could count the fires started by a single main line passing train, and not a logging train at that! While timber operators might close down their own railroads, the main lines would keep operating. Another major group also could be blamed for setting many forest fires: the public. Most everyone was careless, believing that the eternal dampness of this Pacific shore would take care of the flames. The trouble was that this amounted to little more than legend, an old wives' tale. Even among people who should know better, having lived their lives in the region, there was a marked lack of understanding of the dangers.

It would take what advertising people today call the hard sell, getting the message across to the general public. At the outset it sounded like the timber people were trying to pass the buck. Time and again the annual totals on forest fire causes showed where the source of most of them lay. But people merely shrugged their shoulders and wouldn't buy it; they kept right on cursing the timber industry. Meanwhile, loggers and lumbermen who were now starting to accept on a broad scale the philosophy that "timber is a crop" were becoming increasingly fed up with always being the scapegoats for all forest fires when statistics told a different story. There was need for a hard-hitting program to set the record straight and also educate the public as to its own responsibilities toward the region's timber.

Bill Greeley thought the public could be reached. Like most of those in lumbering and forestry, the colonel held a great admiration for and faith in the printed word. The voluminous published reports, studies, surveys, magazine articles, and books on the industry fall short only of those of the Pentagon. Traditionally, loggers were great readers.

Greeley called for ideas from among the membership of the West Coast Lumbermen's Association and also from the able Seattle writer James Stevens, who was steeped in the legends and problems of the woods, and wrote in a straightforward, down-to-earth style. It appeared to be a matter of public education, but how? The answer seemed to lie among the children and youth of the region. Youth, loving "causes," often reacts with enthusiasm. There had been some terrible fires in British Columbia, and in a desperate effort, the Canadian Forestry Association launched a successful fire-prevention program through its young people, who became Junior Forest Wardens. The youngsters were very conscientious, and through them the adults were reached. Why not attempt something similar, asked Greeley, in Oregon and Washington State?

In 1939, the year of the second great Tillamook disaster, Greeley and his associates created an organization with a cumbersome title, "Junior Forest Council of the Douglas Fir Region." The council's structure, as outlined in detail by Stevens, involved the American Legion posts, which were very active in youth work through baseball teams and leagues, Boy Scout troops, and other similar clubs. The two state foresters, Ted Goodyear of Washington and J. W. Ferguson of Oregon, were named cochairmen. The board of directors included Legion commanders, lumber and logging people, and foresters from state and federal agencies. The Legion carried the ball to small and large communities, for its activities reached everywhere, while much of the publicity and promotion flowed from the typewriter of Jim Stevens, who was now the WCLA information director.

Still, it was believed that even more should be done to stress fire prevention. When the Western Forestry Conference was held that summer in Portland, stout-hearted males who had daughters as well as sons at home revealed a condition that three decades hence would be known as "women's lib." The girls were becoming mighty vocal and giving their dads a bad time because they couldn't participate in this fire-prevention campaign. Just why, the fathers were asked, was the program limited to boys?

The answer was that no one thought the girls would be interested. Ted Goodyear said he'd take it up with his boss, Governor Clarence Martin, and Roderic Olzendam thought including the girls was a good idea. Martin believed the program should be expanded to include not only the gals but adults as well. A cautious budget of $5,000 was adopted, half coming from the industry and half from the state, and a steering committee was organized, with Olzendam as chairman.

But Governor Martin was bothered by one aspect of the plan.

"Just who is going to do the job?" he asked the group gathered in his office. "Who is to be responsible? We've got to have someone to carry the ball."

Jim Stevens was ready for this. He threw out the name Stewart Holbrook. Hols Holbrook was a "natural," Stevens contended, since he "belonged" to the lumber world, a tough, rugged, barrel-chested, plain-speaking ex-logger who still owned and wore his hobnails, and was at home with the timber world and its people. Holbrook in recent years had emerged from the brush and the logging camps as a top-grade writer with a mounting output of magazine and newspaper articles, and now books. He had most recently created a national best-seller called *Holy Old Mackinaw*, a lively and informative history of the American lumberjack. He'd published another called *Iron Brew*, and was working currently on a third about the Revolutionary War hero, Ethan Allen, from Holbrook's home state of Vermont.

Lumberjack Holbrook had other appealing qualifications, as Stevens saw it. He'd served as a top sergeant with the American Expeditionary Force of 1918 and was, therefore, in this manner of speaking, one of the Legion's own kind. The fellow was well liked and had respect. He was a human dynamo of energy and enthusiasm for any cause in which he believed, and like Colonel Greeley, Holbrook was very distressed by the mounting losses from timber fires. The only trouble was that Stewart was living in the East, off the Harvard University campus, where he was doing research and occasionally guest lecturing. (In later years, Holbrook would jokingly boast that he was "the only logger ever to lecture at Harvard.")

Jim Stevens wasn't certain whether Holbrook could be wooed back to the Northwest, for Holbrook's interests were exceedingly broad and varied, he was riding a crest of success in his writing, and his native roots were deep in the soil of Vermont. There was some doubt whether this proposed crusade would appeal to Stewart, a practical conservationist who saw clearly both sides of the coin and could get along with opposing factions, yet was very aware of what was happening to America's timber reserves and could exhibit great anger over any recklessness causing forest fires, viewing it as sheer stupidity and waste. Stevens was sure that Holbrook was the right man to put the crusade across. Stewart

knew his way around the logging camps and timber towns, from the skid roads
to the high places, and his copy would likely be of the grass-roots kind, so lively
and fresh that editors couldn't turn it down.

"I'll take him," Governor Martin declared.

Stevens couldn't believe his own sales ability. But now Holbrook had to be convinced, and that might not be so easy. The crusty lumberjack could be plenty
gruff, even with good writing friends, and nobody could predict how he might
react to anything. Stevens agreed to tackle him, when nobody else seemed anxious
to do so; at least there was the common ground of their work about the timber
world, and therefore a healthy respect for each other. But Jim's own anxiety
nearly brought disaster. The thing was on his mind that very evening. He couldn't
sleep, so he decided to call Holbrook in Cambridge and settle the matter. After
several rings and a long wait, a sleepy voice answered the phone. Stevens explained that he wished to telephone Stewart before bedtime to tell him the good
news. The wire sputtered and crackled.

"Bedtime hell!" Holbrook roared. "It's nigh two o'clock in the morning back
here!"

Stevens had forgotten the time difference. He'd blown it for sure, for Holbrook
would likely be sore enough to turn down the offer. His own stupidity might cost
the movement Holbrook's great talent and energy. But finally the author calmed
down and listened attentively to what his friend was saying. Stevens detected a
ray of hope, since Holbrook seemed actually flattered at being approached. The
offer was made at the right time. He'd finished his book and was missing the
Pacific Northwest, so he agreed to return to Washington to fight forest fires with
his typewriter as the first director of the Keep Washington Green movement,
the first such chapter in the nation.

The name, which now has widespread usage in many states and even a national campaign, is of uncertain origin. "Junior Forest Council of the Douglas
Fir Region" wouldn't do at all, even in today's utter fascination with title letter
combinations, for the letters JFCDFR were unpronounceable. "Junior Forest,
etc." was also too cumbersome. Something short, catchy, and on the mark was
needed. Olzendam suggested "Keep Washington Green." Whether it originated
with him doesn't really matter, except to guys like me who are intrigued by the
beginnings of things. Years later in tracking down the name's origin, James
Stevens discovered a small booklet on timber taxation published in 1925 by Forester E. P. Cheney of Minnesota, entitled *Keep Minnesota Green*. The term was
also used by Holbrook's home state as "Keep Vermont Green"; he might have
borrowed it, then planted it in the minds of others, as writers and publicists
sometimes do. And Jim Stevens, too, could have provided the seed when he
penned a parody on "Green Grow the Lilacs," which Ivar Haglund, Seattle folk
singer and logger's son, sang over the radio: "All nature is praying, keep our
forests green. . . ." In any case, the three words caught hold, sparked the public
imagination, and spread across the continent.

Holbrook's deep feeling of concern for the region and an originality of approach launched the nation's first Keep Green chapter in a style that made it
the pace setter. Jim Stevens was correct in his selection of a director, for Holbrook
swung into action with a mighty swath of voluminous material in newspaper
stories and features, and radio commentaries that were done professionally, yet

with the unique terminology and flavor of the forests and its people, which was readily acceptable with few alterations by the state's tough news editors. It wasn't difficult to "sell" forest-fire protection; everyone was against fire. But poorly written material, with no respect for deadlines or sense of timing, would have been placed in the circular files. Holbrook's stories of going forest fires, which he often covered personally, were hard-hitting, factual, honest accounts, gratefully accepted by editors who didn't have staff members to dispatch on such assignments. Pictures accompanied the reports, which often made the front pages.

Stewart wrote radio shows and went on the air himself, for he was a professional actor. He talked to clubs, schools, and service organizations, and toured the lumber towns and camps to speak about fires and Keep Green to cynical loggers and their families. These people accepted him as one of them who spoke their language, unadorned and unpretentiously. He struck a common chord, doing a plain job in a plain and practical way without ballyhoo or window dressing. When he talked of fire being everyone's responsibility, his words had impact and made good sense to the lumberjacks. The aim was to make *everyone* fire conscious, from logger to city dweller who only invaded the woods on weekends.

The Stewart Holbrook System for Stopping Forest Fires Before They Start began paying dividends. The timber beasts were utterly amazed. It brought enthusiastic support from lumber companies and the associations. There were billboards, posters, thousands of pledge cards, and stories and editorials in the state's dailies and weeklies at the rate of 700 to 1,000 column inches per week. By mid-July 160 editorials were published on the Keep Green movement. Forest fires began receiving greater coverage than ever before; even small fires made front pages in a region where anything short of a holocaust was normally buried on the back pages, fires being so common to this timber country that they weren't considered worthy news. A case in point was the *Seattle Post-Intelligencer*, where one fire that might normally have received only a few lines was given earthshaking attention—two stories by the lumberjack journalist, rounded out with pictures, editorial comment and cartoons, all spread across a full page.

The public was aroused, and the statistics spoke for themselves. In 1939 there were 1,516 fires set by campers, sportsmen, berry pickers, and smokers, which blackened 20,716 acres for a property loss of $27,252. There were more fires in 1940, when KWG was organized, but the burned acreage was reduced to 5,726 acres and the loss to $1,835. More amazing was the reduction of burned acreage of all fires, from 103,139 in 1939 to 40,633 the following summer, and the average acreage per fire from 68 to 21.7 acres. In 1941 the fire total stood at 1,562 and then dropped to 842 fires for the summer of 1942. The fires were smaller, too, as workers and the public became alerted to the true dangers and quickly took action, rather than bypassing a small fire for someone else to tackle.

Much of the Keep Green message came across to the adults through their offspring. It was paying huge dividends, and there was the first hope in generations that eliminating fire from the forests—or at least reducing the destruction to a low minimum—might at long last be within reach. As things developed, the timing was also more than right. Following Pearl Harbor, the program took on additional prime significance in the need for protection not only from carelessness but from arson and fires started by Japanese incendiary bombs launched from offshore submarines and later long-range balloons. The war meant increased public responsibility to keep fire out of the woods and also rangelands east of the

Cascades, for fire meant not only a loss of timber but a loss of time and labor in a critical wartime period when labor grew increasingly short.

Stewart Holbrook worked for the crusade for four summers at a salary and expenses of $2,700 for five months. That gave him winters to devote to his other writing. His experiences with the Keep Green movement inspired another book, *Burning an Empire,* a history of American forest fires that had national impact, with its introduction by Colonel Greeley and dedicated to his friend who had once gotten him out of bed to talk about Keep Washington Green: James Stevens.

Not to be outdone, Oregon was quickly clamoring for a Keep Green program like that in Washington. There was a certain rivalry between the two states; if Washington were to be kept green, Oregon must be that much greener. A lumberman of note, Edmund Hayes, urged the launching of a similar crusade and gained the sympathies of Nelson S. Rogers, the new state forester who was a native of the Tillamook Burn country and therefore had the world's finest example of what fire could do to prime woodland on his home doorstep. It was Rogers' foremost ambition while state forester to see that the Tillamook country be brought back to life, a task of such momentous dimensions in both human effort and financing that foresters and lumbermen scoffed cynically at the suggestion.

Colonel Greeley's lumbermen association provided the impetus behind both the original Keep Green programs, assisting with the organizing and much of the publicity, promotion, and ideas of what forms the crusade could take. But Greeley saw the crusade as "much more than a summer ballyhoo."

"The KOG and KWG campaigns should grow into permanent means of public education," Greeley declared. "They have an important part in our industry program of forest conservation."

The Oregon movement borrowed from Washington, then tailored things to suit the Beaver state. One early proposal of the first director, Richard Kuehner, stemmed from his outstanding work with 4-H youngsters in the creation of a Green Guard auxiliary to the adult campaign. The Green Guard plan was adaptable to any of the youth groups such as 4-H, Girl and Boy Scouts, Camp Fire Girls, Future Farmers, churches, schools, and community service organizations in small towns, whistle stops, and cities. Kuehner developed a Green Guard kit for every applicant, containing membership card, manual on fire protection, arm band, gummed poster for the youngster's home, and other items that would excite and inspire the kids.

Following the first announcement, the Keep Oregon Green office was swamped with requests totaling around 20,000. Buried under reams of paper, Kuehner desperately ordered his press director, Arthur W. Priaulx, "For Heaven's sake, don't send any more publicity stories out on the Green Guards. We'll soon go broke trying to pay our way through this deluge of requests."

Indeed, the youth phase of the Keep Green movement became so popular that it required little promotion, for the word just got around. When Kuehner left the program for World War II service, his parting shot at Priaulx was: "If you're smart, you'll keep quiet about the Green Guard activity and let the kids pass it along by word of mouth."

It was sound advice; in the coming years the only "organizational" mention of the Green Guards came in press releases urging youngsters to enter poster or essay contests, and when there were individual acts of service or heroism. Over three decades, thousands of youngsters have been proud members of the Green Guards. By 1972 the total reached 183,403 children on membership rolls, now into the second generation. Youngsters took their memberships seriously in the dousing of abandoned camp and warming fires, and the reprimanding of adults for carelessness with their cigarettes. Through the Green Guards the fire message was relayed to parents, grandparents, aunts, uncles, cousins, and friends. Furthermore, the training stayed with the youngsters into adulthood and with many of them throughout their lives, thus accumulating over the years a much greater public awareness of any kind of fire danger in the woods, rangelands, and even in the cities. There was a mounting concern, too, over the general importance of conservation and the preserving of wildlife long before the great ecology movement of the 1970s.

During the years of World War II, youngsters felt they were contributing to the winning of the war by urging fire precautions on the homefront. They took their patriotic responsibilities seriously; it wasn't play with them. One black night five Green Guards wearing their "official" arm bands appeared at the main gate of the huge Oregon shipyards of Henry Kaiser in Portland, where fleets of Liberty and Victory vessels were being turned out for the war. The Green Guard captain displayed his credentials, and all five looked determined. The Guard captain announced that they'd been commissioned by the governor of Oregon to locate all fire hazards and, therefore, they wanted to inspect the yards. The man at the gate, helpless in the situation, called his superior. The boys hung tough in their demand. It took considerable diplomacy on the part of shipyard officials, including an emergency call to the state capital, to sidestep a crisis without any loss of face or dignity to the serious-minded Green Guard contingent.

The Washington crusade once sustained a similar crisis with its youth during an annual poster contest on fire prevention, which had from 15,000 to 20,000 submissions from first through twelfth grades and cash awards to the winners. The posters were stored in a third-floor room of Anderson Hall, the Keep Green headquarters on the University of Washington campus. Just before judging began, a forestry professor, seeking classroom space, tossed them into a bin bound for the university garbage dump. The horrifying discovery created a condition second only to a major forest fire. For two anxious days and sleepless nights, KWG Director Bob Lyman and his assistant, Bill Massey, conducted a fragrant yet distasteful search through the trash and swill of the huge dump with a carefully operated bulldozer, two other attendants, and an irritated covey of hundreds of seagulls before the posters were located and the troubled Lyman could forget what was foremost in his mind: how he was going to explain to all those loyal youngsters.

The Oregon program didn't have Stewart Holbrook during its formative years, nor anyone of his particular caliber, although there was much exchange of ideas and information between the two chapters, along with spirited rivalry, since both were backed by the industry. Holbrook also recognized no borders, giving of his talents whenever possible to spread the word on fire prevention. Once he stacked up traffic for a great distance on both sides of the Columbia River with a border meeting of the two governors midway on the lone span of the main north–south highway, so that in special ceremonies they could pledge mutual assistance in fire emergencies. It was a bold stroke, which created many angry motorists, but Holbrook got what he wanted: headlines galore, plus a massive traffic jam in the name of Keep Green, which delighted him immensely, since it disrupted the normal flow in the affairs of men.

It took a certain kind of individual—personable, full of enthusiasm, overblown with ideas and the confidence to put them into action, plus a strong belief in what he was doing—to make Keep Green work at its fullest potential. As a pacemaker, Holbrook had been a fortunate choice for the first chapter in Washington. Oregon searched for a similar leader, and in 1948 found him in a lanky forester with a wit and a way who made a second career out of the crusade and became Mr. Keep Green himself.

Albert K. Wiesendanger, a native Oregonian, had been a pioneer career forester with Uncle Sam since 16 years of age. He was also spreading the message of forestry and fire protection among young people in the Northwest long before it was recognized as anything beyond minor significance to the over-all picture of lumbering and protecting the Big Woods. In 1909, when Wiesendanger passed his first civil service examination to become a "messenger" with the U. S. Forest Service, foresters were generally looked down upon as "bug men" and "pismire superintendents" by people of the tall and uncut. By 1916 Wiesendanger had passed other tests qualifying him as a forest ranger and was assigned to the Oregon (later Mount Hood) National Forest at the Eagle Creek campgrounds in the Columbia Gorge. Even in those times foresters were embarked upon a program of educating the public, especially the young, in the importance of the woods, its wonders, and the need for protection against fires. Much of this was accomplished in the off-seasons of fall and winter by traveling around to schools and communities. One of the other rangers was doing this work and invited Wiesendanger to come along to operate the slide projector. Albert noticed quickly that the fellow's talks were too technical, exceedingly dull, and often going well over the heads of the youngsters.

Wiesendanger felt he could do a better job by lightening the program and amusing the children. When the opportunity arose, he asked if he might borrow the slide projector. His first effort was so successful that he kept right on doing it, finding he enjoyed this kind of activity. There's a bit of the vaudeville ham in Wiesendanger, who entertained while teaching, often playing his harmonica to amuse the kids. Then he would announce, "I'm going to play this thing backward." The kids leaned forward expectantly, doubting that he could; then Wiesendanger would turn around, back facing the audience while he played, and bringing hoots and yowls from his young listeners.

Each winter Albert and his good wife Mickey moved to Portland, and he began making the rounds of schools in northwestern Oregon and southwestern Washington. He appeared in an authentic ranger's uniform of that day, with leather puttees, choke bore pants, and peaked campaign hat, an outfit similar to uniforms worn by the doughboys of World War I. He brought along a fascinating array of equipment, including cooking utensils, fire tools, pack saddles, compass, maps, and an intriguing instrument called a heliograph, for flashing signals from one mountaintop to another. The youngsters were enthralled. There was no telling how many boys Albert steered into forestry as a result of his displays, photographs, and colorful tales of the woods, probably the first public

31. The great Tillamook fire, and subsequent jinxlike holocausts every six years, left a 300,000-acre wasteland in northwestern Oregon. Its black snags of once-fine timber symbolized what became world-known as the Tillamook Burn. —*Author's collection.* (Chapter IX)

32. Mile after rugged mile of the Tillamook Burn bristled for decades with snags of once-grand timber that would have lasted for generations. The snags created a tinderbox of future fires so that the land couldn't come back. —*Oregon State Forestry Department*. (Chapter IX)

33. The Tillamook Burn as it appeared in the mid-1940s. The land, good for little but growing timber, wouldn't reforest naturally, and while a delight for deer hunters, it was an irritant to the public and foresters. —*Author's collection.* (Chapter XII)

34. The salvage effort in the Tillamook Burn was tremendous, encouraged by high World War II and postwar prices. Even "portable" sawmills like this worked over the small stuff. —*Ellis Lucia photo.* (Chapter XII)

35. Timber companies merged their efforts during World War II to mine salvage logs of the Tillamook Burn for the war effort. Millions of board feet were
funneled through the roaring Consolidated Timber camp at Glenwood. The
Lyda fire of the original Tillamook holocaust broke out about a mile up the
canyon to the far right. —*Courtesy A. C. Johannesen.* (Chapter XII)

36. Many sound trees, even when charred black, were found in the Tillamook Burn. This salvage operation was dangerous railroad-incline logging. Loaded cars were eased down the steep grade by cable (lower left) and braked by hand. —*Ellis Lucia photo.* (Chapter XII)

37. Long pole trains of salvaged logs rumbled through the Burn to mills in the Tualatin Valley and Portland. The total salvage value eventually equaled the original loss estimate. —*Ellis Lucia photo.* (Chapter XII)

38. The Owl Camp, later renamed Rogers Camp, in the heart of the Burn, as it appeared in 1948, when it became field center for experimental projects to replant the great disaster area. —*Ellis Lucia photo.* (Chapter XII)

39. Ed Schroeder, a career forester in Oregon, accepted the challenge of licking the "six-year jinx" to reforest the Tillamook Burn. —*Author's collection*. (Chapter XII)

40. Innumerable public meetings were held to determine what to do about the Burn. This informal session was held in 1948 at the Owl Camp. Ed Schroeder is seen facing camera in center, with Rudy Kallander, state rehabilitation forester, to the right. —*Ellis Lucia photo*. (Chapter XII)

relations man the U. S. Forest Service ever had, certainly in the Pacific North-west.

However, it became very apparent that this was Wiesendanger's particular line, where he could make a strong contribution to the controversial forestry program, which was feeling its way with federal and state agencies. He was promoted to recreation ranger in 1931, assigned to Mount Baker National Forest at Bellingham, Washington, at Heather Meadows near Mount Baker Lodge. The lodge was destroyed by fire in 1934, and Wiesendanger was transferred back to Portland and then to Timberline Lodge, the massive snowbound public edifice constructed on the slopes of Mount Hood. Once again he was recreation ranger until the lodge was closed down by World War II. Then he was directed to Columbia Gorge headquarters at Cascade Locks near Eagle Creek, and Wiesendanger felt he'd come full circle. Yet all the while, he kept on making the school circuits during the winters to educate students about the forests, and by now he was into the second generation of children whose parents had been his audience years before.

Wiesendanger's love of the forests and his love of people, especially youngsters, his energetic and dynamic way before an audience, and his long success with the Forest Service brought him to the attention of the Keep Oregon Green directors. The program was about six years old in 1948 and was losing its second director, or executive secretary, Charles Ogle from Klamath Falls. Ogle was being urged to resign to become a lobbyist for the industry in the Oregon legislature, for he was well-versed on forestry matters. Ogle, who often visited Wiesendanger when he came his way, was asked to suggest a successor and, impressed with Al's experience in this business of public education, submitted his name. The board picked him from about a dozen other candidates for the job. But Wiesendanger wanted to stay with the Forest Service until his retirement, not too many years away, since he was now 55.

Convinced that he could be of service in the Keep Green position, he took an optional early retirement to become KOG executive secretary, headquartered in a former CCC building at state forestry headquarters in Salem. Wiesendanger proved quickly to have the bounce and boundless energy of the Stewart Holbrook pattern; it wasn't long before the KOG movement was confronting everyone in Oregon through the press, radio, billboards, posters, window cards, Green Guard activities, and in 1,001 different ways. Wiesendanger really got around, not only to schools but in the woods, mills, county and state fairs, community celebrations, the Portland Rose Festival. parades, and rodeos, and he was likely to bob up anywhere there was a handful of people. Keep Oregon Green signs and posters smacked you in the face everywhere you looked; even far in the back country, rounding the bend of a trail, you'd come face-to-face with a KOG reminder. He became readily identified as Mr. Keep Green in building one of the most active and effective programs in the nation; and as the years passed, the dean of the Keep Green movement grew into a living legend. He had impressed the warnings of forest fires on three generations of Oregonians, and, like Holbrook, set the pace for such movements in other states, and nationally, to combat one of the country's most dread disasters. He received numerous citations, among them the second "Silver Smokey" award by the Co-operative Forest Fire Prevention Campaign—Advertising Council, Inc., National Association of State Foresters, and the U. S. Forest Service—and the Western

Forestry Award of the Western Forestry and Conservation Association. In 1974 at the age of 82, Wiesendanger still hadn't taken his "early retirement," but was going strong as ever, now with a budget of $40,000, mostly from the industry and private donations, and including a $7,500 sum from the State Forestry Department, but no federal money and the same staff of two, himself and his secretary, and unofficially, his good wife.

Wiesendanger never misses a trick in promoting Keep Oregon Green, and other directors have taken his cues. Over the years KOG has developed a broad variety of brochures, informational bulletins, films, folders, and gimmicks to keep the KOG reminder before the public, constantly, the year 'round. There are souvenir ash trays for conventions, pocket knives, key rings, and forest-fire prevention kits for schoolteachers. Some materials are furnished through the U. S. Forest Service Smokey the Bear program. Wiesendanger even devised a metal Keep Oregon Green sign to be affixed permanently to wooden panel bulletin boards posted along trails and in forest camps. When the summer season is over, the paper notices come off, and the Keep Green sign stays without weathering all winter.

"Keeps folks from tearing down the bulletin boards for firewood and serves as another reminder," Wiesendanger explains.

Wiesendanger has readily exchanged ideas, brochures, and publicity materials with other Keep Green chapters and other conservation movements around the country. His enthusiasm and many years in the work have made him a national influence on the crusade, which has spread to some 40 states, Keep Alaska Green being the latest. The crusade shows no sign of lessening; California, for instance, has at least three separate Keep Green programs. The catchy slogan also was adopted for a Keep America Green campaign, and borrowed in other ways. In the early 1970s, at the suggestion of Governor Tom McCall, Oregon added to the original slogan with "Keep Oregon Green and *Clean*." The parent Northwest chapters have always been alert to the shape of things, and sometimes were out ahead.

In an age ridden with power intrigue and maneuvering of special interests, the Keep Green crusade kept its appeal by being free of complicated organization, politics, and profit-taking. Anyone could do his own thing, on his own good time. There are no regional or national conventions of Keep Green directors and executive secretaries. It has been kept simple and straight, and wise directors like Al Wiesendanger give controversial matters such as clear-cutting a wide berth. And despite the many organizations that have sprung up in recent years devoted to concern for the environment, quality of life, and forest and wildlife preservation, the Keep Green movement hasn't diminished in its importance. People need to be constantly reminded, it seems, and the Keep Green movement has taken on new dimensions with an upsurge of population in the Pacific Northwest, even to warning people bunched in crowded suburbs about the fire dangers in vacant lots and woodlots. The fire education of youngsters and adults is a continuing one. Also, thousands of newcomers to the Northwest timber region have little basic understanding of the ways of the Big Woods and are likely to be far more careless than the natives.

Great numbers of people from the growing urban centers are escaping into the forests; and the Pacific Northwest is very near for some 20,000,000 Californians who find it a handy summer playland, just up the freeways a few hundred miles. In Washington State in 1916–25, with a population of 1,356,621, there were averaged 1,136 man-caused fires, or 838 per million population. By the period from 1956–65, Director Ed Loners points out, the average was 394 fires per year blamed on careless humans. But with population climbing to 3,340,000 and teeming urban millions using the forests and recreation areas, the figure shot up to 526 fires per million per year in 1966–70, and 729 for 1970. There were 2,436 man-caused fires in Washington State that year, the thirtieth anniversary of the first Keep Green program. Oregon was trying to hold the line under similar conditions of the influx of new people and a mounting tourist industry, and was able to lower the 1970 total of 2,014 man-caused fires down to 1,224 in 1971. While it seemed certain that even more intensified campaigns were needed, one can only imagine what the fire tolls might have been without the Keep Green crusade; for while much new modern fire-fighting equipment and techniques were hitting outbreaks hard, the threat from man's own carelessness, ignorance, and lack of caution would likely boost the totals many times. Colonel Greeley's battle cry to "do something about fire" remains valid even in an age of ecology and concern over the environment, so that Keep Green continues to hold a key place in the strategy of education and protection. Director Loners summed it up in a few words:

"Fire prevention in the forests and on the rangelands is a *constant* job."

To which the dean of the Keep Green movement in his eighty-second year would add his own "Amen."

Chapter XII

Licking the "Six-Year Jinx"

It is hoped that Oregon conservatism does not show itself at the suggestion of a $25,000,000 limit, which those who have made a study believe offers a sound business investment of the long-term type which the state can assume. It is also to be desired that the legislature will not show its customary disregard of the carefully prepared recommendations which are often filed for its consideration by such groups as interim committees, and disregard the suggestions of the special forestry committee. The Tillamook Burn management has much to do with determining the economic future of this area. Forest-fire efforts have been a repetition of too little, too late. Before there is further dissipation of this natural resource so important to this area in particular and to the state in general, it is time for the state to inaugurate a real program, which will not play "penny ante" with a problem of large proportions as well as great opportunities.

Editorial
Washington County News-Times
November 14, 1946

By the summer of 1945 the bloody tragedy of World War II had about run its course. Germany surrendered, and now the pressure was hard upon the Japanese in the Pacific Theater, and against the full weight of American power, it was hoped that this sorry chapter of human agony might come to close before the end of the year. Already attention was turning to the long-neglected internal matters of the United States.

Yet in this hot, dry summer, looking toward peace, the Tillamook Burn held to its "six-year jinx," as loggers called it, and exploded once again with renewed ferocity. It was another fire that couldn't be stopped in this combustible corner of Oregon; the smoke clouds mushroomed once more into the sky to be seen by the densely populated city of Portland and by vacationing thousands on the Oregon seashore. The holocaust this time was fought by high school students, for manpower was still short. More old growth was destroyed, some 30,000 acres;

and over 200,000 blazing acres and the young trees trying to struggle for life were added to the five-county desolation that had now existed for over a decade and might well be there for at least a century.

"Why can't *they* do something?" was the public outcry. The fact that thousands could easily see the fire burning and the desolation firsthand cannot be overemphasized. This was the most important factor in what became the major turning point for forestry in the Pacific Northwest. The Tillamook Burn was hardly something you could sweep under the rug, or for which you could make excuses. It was right there on your doorstep, and this chain of infernos was a constant reminder of how man had ruined his world.

Still, there were glimmerings of hope. One of the good things of World War II was the unbelievable salvage effort that took place in the Tillamook Burn. Much of the fire-ravaged timber was found to be sound and disease-free, but under normal logging practices, impractical to take from the standpoint of economics. Oh, there had been some salvage logging since the original explosion of 1933, for hungry lumberjacks moved back into the territory before the embers were hardly cool. It was the strangest logging operation in history, felling and bucking black timber in soot and ash up to their knees, and none of the familiar signs of a living forest about them. Emerging from a logging day, they were covered with ash and soot, black from hair to hobnail, resembling miners from the coal pits. People called them "Tillamook coal miners." The market was unpredictable from the Depression and because sawmill operators were doubtful about these great black logs at first, until discovering that the interiors contained, of all wonders, thousands of board feet of sound lumber.

While the gyppos worked on a shoestring gamble, other outfits—among them Blodgett, and Crossett Western, which were hard-hit by the original fire—merged into a single firm, the Consolidated Timber Company, for the sole purpose of logging The Burn. It would be no gamble at marketing the cut, for Consolidated had its own mills. A deal was also made with Tillamook County to eliminate the usual property taxes and instead, loggers paid $.05 per 1,000 feet of timber logged and sold. Still, it was a slow process, for the price for logs was down. Then, wartime inflation and the sudden high demand for lumber for the war effort, including the great shipyards and government housing projects, changed all that. The necessary profits and the demand were there, and it made good sense to get out that black stuff for war. Hundreds of log trucks, one every five minutes, and the great pole trains of the twisting PR&N railroad streamed through the little towns to the sawmills with their odd cargo—blackened and scorched logs of the Tillamook, timbers that would have been rejected out of hand and allowed to rot or become infested with disease while fresh, living trees were felled for that most wasteful of human endeavors, war.

Consolidated Timber, under the hard-headed direction of Lloyd R. Crosby, kin of the famous crooner, and Alf C. Johannesen of the old Crossett Western outfit, even won the coveted Army-Navy "E" banner for its outstanding job of wartime production "to beat the Nazis and the Japs." The volume was tremendous, extending into the postwar housing boom, when the pattern of logging changed from steam donkeys and locomotives to gasoline and diesel, and when power saws were invading the operations for the first time. The demand for housing lumber mounted, so that there was value in lesser grades, and many small independents were scrounging for stumpage in The Burn, with just about every kind of contraption imaginable, including the first portable sawmills moved on truck beds from one contract area to another. The number of logging out-fits rose from around 50 to over 200 in those postwar years, and that didn't in-clude the reloggers, who took just about anything in the way of a windfall or snag that was anywhere nearly "sound." In the end, the Tillamook Burn was close to the "full use" of the residue the timber industry came to regard as es-sential two decades later, and therefore was ahead of its time, demonstrating that it could be done if the need were there. And in all, the great salvage opera-tion totaled The Burn's original estimated loss of 13,000,000,000 board feet of lumber, as the biggest single salvage effort in the long history of the industry.

There were other benefits. The salvage effort cleaned out many of the bleach-ing snags that marched by the thousands along the ridgetops and through the dark canyons, and thereby lessened the fire hazard. Much of the region's ex-plosive potential was blamed on these snags, which not only were magnets to lightning, but were also a means by which fire could travel at high speed for miles, leaping from snag to snag, ridgetop to ridgetop, making fires impossible to control. The salvage loggers, although they may not have realized it, were making The Burn safer and helped to clean it up—which would be necessary if ever there were to be any kind of effort to return this lonesome land to produc-tivity.

There was talk of this now, but little more than talk, for it seemed the im-possible dream. Nothing even resembling it had ever been attempted. The area was far too huge, too rugged, too ravaged by fire, too choked with combustible debris, and the necessary work to rehabilitate and then protect any new forest was too staggering to consider. Moreover, the cost would likely be prohibitive and the investment beyond the lifetimes of those who would do the work and provide the financing, since it would likely take a century or more before the Tillamook Burn could be productive again.

"The simplest approach is to regard the entire Burn as being beyond justifi-able human effort, to surround it with a superfireline and let nature take its course, whether it takes 100 or 300 years to grow a new forest," declared a U. S. Forest Service report of October 1945. "While such an approach is extremely pessimistic, it is much sounder than a continuance of protection expenditures at the present level."

Let nature take its course . . . albeit, let the damned thing burn! Nelson S. Rogers, the Oregon state forester, refused to accept that approach. He found a few supporters among the forestry and timber fraternity, and more among the people. Rogers was prejudiced about that country. He came from the Vernonia and Forest Grove area, and could well remember those beautiful forests of his boyhood, mile after country mile of lush green timber that was as fine as any in the world . . . the thick undergrowth that gave shelter to deer and mountain lion and elk and bobcat, and the countless smaller creatures that make all life amusing and wonderful . . . the gushing, bounding, racing streams alive with salmon, steelhead, and native trout, hiding in deep pools or pushing their ways to their home spawning grounds. It was a paradise of extreme rarity, as Rogers remembered it, and although he was too far down the road ever to see it return to its former grandeur, he was determined that while still state forester, some kind of a beginning might be made that would eventually bring back the tall timber.

The 1945 fire roaring through the Salmonberry country and the end of the war provided the necessary impetuses. The first in a series of meetings, which went on all winter, was held in the shadow of the Big Burn at Forest Grove to consider what might be done. About thirty city, county, and state officials, foresters, lumber people, and a smattering of interested citizens turned out for the lengthy session, chaired by a local state senator, Paul L. Patterson, who would later become governor. The great concern was that the years were slipping by and that erosion plus The Burn's combustibility made this probably the last chance to bring back the forest. The erosion was tremendous without foliage to hold the soil, and Tillamook Bay was rapidly filling. Every new fire enhanced the problem, for protection through the private association, with its limited and ofttimes depleted funds, was nil, and efforts at reforestation were haphazard at best. Nature couldn't do the job, unless given many centuries, and by then the complexion of this vast timber region might well be completely altered.

"We are just a little burned up at the way the program has been handled in the past," declared Judge Harland M. Wood of Tillamook County at the Forest Grove meeting. "We don't think it has been handled. We're going to have a fire in another few years because nothing has been done about it, and nothing is going to be done about it."

What to do with The Burn was being tossed about like a medicine ball, asserted one newspaper scribbler. Rehabilitating some 2,500 square miles of topsy-turvy land, as explosive as a military ammunition dump and as infested with disease and debris as a garbage pit, boggled the mind. It involved complete fire protection, dealing with the extensive erosion problem, and in the end the tedious task of planting millions of trees, mainly by hand, although the helicopter offered doubtful possibilities, for in 1945 the helicopter was untried for forestry work. Perhaps, suggested some at the Forest Grove hearing, much of The Burn should be seeded to grass, turning it into a huge grazing land for sheep and cattle. Grasslands would at least halt the erosion. That brought outcries of objection. This land was too valuable for grass, too steep for grazing other than by mountain goats. It was good for nothing but growing timber, but at that, it was the finest in the world.

Nels Rogers certainly wasn't willing to settle for grasslands. His department, at his prodding, already had drawn up plans for rehabilitating The Burn, including the salvage work, clearing and cutting snags, classification of the lands and mapping them, construction of firebreaks, roads, and trails, rodent control, experimental seeding, and then an all-out replanting effort, at an estimated cost of $18,000,000. There was an urgency, Rogers stressed, for each year lessened their chances of success. It should be done as fast as possible, for time was running out. But $18,000,000 was a sizable sum for that time, and for a state like Oregon. It was doubtful that Oregonians would approve such an amount, yet there was a reluctance, from home pride and a desire to sidestep federal interference, to approach Uncle Sam for financing.

Through various public agencies—the State Forestry Department, Soil Conservation Service, the county courts, the cities and towns, the U. S. Forest Service, and private landowners—a sum of $75,000 had been raised for a starter experimental project that very fall in both grass and tree seeding in a one-mile strip in 13 sections of the Hillsboro and Forest Grove watersheds. The aim was to discover whether such seedling was practical. It was a token test, but it was at least a beginning. And that winter, too, 250,000 Douglas fir and Port

Orford cedar seedlings were set out in the vicinity of Cochran and the old Reeher's Camp, not far from where the original fires took off.

Meanwhile, hearings continued around the state in an attempt to devise some long-range plan for licking the problem of The Burn, which by now was becoming internationally known as the scene of constant forest-fire devastation. Fire remained the leading concern; there seemed little use in pouring huge sums of public and private money down a rat hole. Until that problem was licked, replanting was ridiculous. And those thousands of snags weren't merely a public eyesore. They made fire control impossible. In the inferno of the 1945 fire, 2,000 men were on the lines, and for the first time an unlimited force of equipment was brought in by the state and on loan from the U. S. Forest Service and private sources. Time and again the flames were halted on ground where snags were scarce, yet where the snags were thick, fire leaped the lines to go on the rampage into unburned areas, ravaging new growth, and beyond control until the autumn rains fell.

The snags must come down and the task would be tremendous, since they were as thick as porcupine needles. It would also be expensive, and there was much public misunderstanding on the need for the job. Why couldn't they just move in and plant the trees? But as prewar lumberjacks and then Consolidated Timber had learned, the snags were of marketable value. The plan was to let tracts out for bids to logging operators wherever public lands were involved.

However, the snag problem wasn't simply a matter of cutting them down. While wartime salvage logging built many roads and cleared sections of fire-kill timber, haphazard logging was now blocking chances of any large-scale reforestation program. An area might be logged over four or five times for different qualities of sticks, beginning with high-grade peelers and sawlogs, and then dropping down to poorer stuff handled by another operator. Since 1940 the state had been quietly acquiring Tillamook Burn lands through allowing landowners to deed property to the counties and withhold payment of taxes. This didn't include the timber, and the $.05 per 1,000 feet was payable at harvest time. In other situations, foreclosures were made and the lands committed to long-term timber sales contracts with the counties. Some contracts were assigned for indefinite periods, while other holders exercised options of renewal. Now in this postwar period, with its housing boom, inflated prices, and lumber shortage, new markets were developing, especially in the areas of the lesser grades and pulp wood, both a hint of things to come. Thus contract holders were slow to give up their holdings for fear of losing something of value in additional salvageable material in the next few years. This reluctance plus the continued widespread logging throughout The Burn country would hamper the rehabilitation project during its first half-dozen years.

There was a crying need, too, to tighten up on the loggers. Better fire regulations and quick shutdowns in critical weather were necessary, for as reforestation became a reality—and men like Nels Rogers and publisher Hugh McGilvra of the influential *Washington County News-Times* at Forest Grove were certain it was only a matter of time—protection would become all-important. The Northwest Oregon Forest Protective Association, one of the state's oldest, was broke. Chief warden Cecil Kyle revealed that it cost some $300,000 to fight the 1945 blowup. This was, of course, one of the major weaknesses of the private associations, for wardens were inclined to hold back on hitting a fire hard in the beginning because of potential cost, risking having it run wild, resulting in a far greater loss of stumpage and spilling over onto adjacent timber properties.

In my own years as editor of the Forest Grove weekly, I developed a "sense of smell," so that by stepping out the door of the newspaper office and scanning the sky, I could detect a fire showing up over the hills. But a call to Warden Kyle at Forest Grove headquarters would likely get a quick denial, or an expression of surprise.

"We got no fire," he would proclaim.

Later, of course, Kyle had to backpeddle as the flames spread and smoke boiled plainly over the hills. Then Kyle, who was a controversial figure, had to admit that the Tillamook Burn was ablaze again, the word spreading rapidly via the wire services along the Pacific Slope and nationally, so that by 10 o'clock that evening, the sorry fact was heard by thousands over the old "Richfield Reporter" newscast from Los Angeles.

That October of 1945, the Oregon State Board of Forestry canceled its contract with the protective association and assumed direct protection control of The Burn. It was the beginning of a new era in Pacific Northwest fire control. The state took over the few buildings and sparse equipment and moved in its own men, headed by John Edward Schroeder, a ruddy-faced, rangy, and friendly career forester who would hand-pick his own top sergeants to make The Burn fire-free, and also launch the significant first stages of what would soon become the biggest tree-planting project in the world. Schroeder, who was only 32, also stepped into the biggest mess ever facing a state or federal forester, not only the snag-infested desert he was supposed to replant, but the bitter feelings that existed throughout The Burn country, with most everyone, from tough loggers to small ranchers, clinging to their meager portions of a largely unproductive land.

As with the U.S. presidency, one wonders why anyone would want the job. Ed Schroeder didn't seek his job out, but he had shot off his mouth on the fire lines that summer, and it is a rule of order, like a Chinese proverb, that he who sticks out his neck usually gets clubbed between the shoulders. Following many weeks of watching that fire go on the rampage again and again, after crews thought it was corralled, and putting up with disorganization, petty jealousies, and a lack of unity and battle planning under the protection district, Schroeder, who normally is mild-mannered and quiet-speaking, blew up like a crown fire.

"By damn," Ed exploded hotly, "if I were running this show, things would be a helluva lot different."

Schroeder forgot what he said, but others didn't, and it got repeated at state headquarters in Salem. After returning to his regular post with the state district at Roseburg, Ed received a telephone call from Nels Rogers. Did he recall what he'd said on the fire lines? No. The statement was repeated verbatim, and then he had a vague recollection. Well, did he mean it? You damned betcha. Then how about it?

Schroeder told Rogers he wouldn't consider taking the job unless he had command. The old protection district must be completely eliminated. This meant sacking Cecil Kyle and his brother Herb, who ran the Jewell forestry station.

The Kyles were well entrenched, having long been in the jobs, and while they had their enemies, they also possessed many friends among the loggers.

"It won't work unless I have full charge," Schroeder declared. "I'm only 32 years old and a stranger to that country. I need to have authority."

It was agreed. So here was Schroeder in charge of the most hopeless forestry district in the nation. He went out alone to look over the country and inspect the facilities. Unable to find a place to stay, he bunked down in the tiny forestry headquarters building outside Forest Grove, where there was a bed in the back room. He could feel the animosity against him everywhere he went, for many independent people in the back country didn't like the state taking over, and especially that Cecil Kyle had been fired.

One evening after a full day in the woods, Schroeder returned to the headquarters building. He found a double-bitted ax lying on the bed. He wondered if it were an omen—a threat—and spent the night worrying about it.

The next day Cecil Kyle located him.

"Say," he asked, "did you find an ax lying on the bed at headquarters?"

"I sure did," Schroeder replied. "I thought it was an omen."

"I was afraid you might," Kyle said apologetically. "I was in there yesterday and used my ax to get a set of antlers off the wall. Laid it on the bunk and forgot it."

Schroeder brought along three able assistants: Curtis Nesheim, a thick-legged, tough forester with the gravelly voice of a top sergeant, to head up the rugged fire protection division; Frank Sargent, a good-natured and able woodsman who grew up in The Burn country and therefore knew it well, to boss the rehabilitation program; and William Phelps, a lanky, studious forestry technician, to handle land management. R. M. "Rudy" Kallander worked from the state headquarters at Salem as research forester in charge of the project, liaison of the 16-point program for state forester Nels Rogers and his deputy, George Spaur, who would succeed Rogers following his death just as his dream project was getting under way. Kallander was a young, university-trained, eager career forester who was well aware of the challenge and the importance of testing every new pioneering idea and technique in what would become the world's biggest forest laboratory. Kallander always insisted that the big project was "forest rehabilitation," not "reforestation" (I, for one, could never separate the two), yet years later after he'd become assistant dean of the Oregon State University forestry school, he described the achievements of the Tillamook Burn as a time "when we were thrashing around in all that dirt and mud and ashes and stuff."

The initial years were ones of feeling the way, as other research foresters joined the team to discover through trial and error the best means of licking this fire-ravaged region and keeping it well protected. Vance L. Morrison, one of the many who followed Ed Schroeder to the Salem headquarters when Ed was appointed state forester, was experimenting with slash disposal methods. Dale Bever was trying different seeds and methods of planting. Soon there were many experimental projects, financed by the research tax, scattered throughout The Burn: hand planting, aerial seeding, fire hazard reduction, natural regeneration, slash and gorse control, forest management and protection, roadside strip clearing, and second-growth studies. District warden Schroeder the organizer, whose pleasant, soft-spoken style gained the most from his men (many would work for him when they wouldn't for anyone else), oversaw the entire project, coordinating its many facets. The program grew larger and more varied, eventually involving not only state foresters but those of the Corvallis forestry school and private industry in this all-out effort to beat the jinx. But it was quickly obvious from the outset that the first obstacle remained that old bugaboo, this tinderbox.

Feelings continued running against the new state regime. In Clatsop County, loggers pelted the foresters with rocks, and so did schoolchildren at Jewell. The harassers were said to be "friends of the Kyles." A house planned for occupancy by one of the wardens was suddenly refused, and it took a lawsuit lasting six months for the state to establish its legal right to the building. Schroeder and his aides got into heated arguments with logging operators who failed to follow orders or to shut down. The lumberjacks weren't used to this kind of authority from a state agency. At times, it came near to physical violence, but Schroeder held his temper and his fists, and managed to sidestep throwing any punches, which could have been fatal for the project. His own husky build plus his knowledge and sympathetic understanding of the private operators and their problems stemmed from solid background as a sawmill man, in his family blood since the day his pioneer grandfather set up a mill in the Willamette Valley. He could talk the language. Yet things remained touch-and-go for a long while, for loggers, timber operators, local officials, businessmen, and the public remained noncommittal. The state foresters were on trial, and they knew it. They realized also that they were living in a storage bin of explosives, where another bad fire might blow the entire project, still to be approved by the state legislature and the Oregon electorate.

"There were times when I felt I didn't have a friend in the world," Ed later reflected. "When I got to feeling low, I'd go up into the Stimson country, where I was during the fire. I did have some friends there. Talking to them would bolster my morale."

Schroeder had his personal plan to whip The Burn, and it involved the lumberjacks. From his own past, he knew them well and had faith in them as reasonable men.

"Educate them," he declared. "Outside lightning, people start forest fires. Most people in the Big Burn are loggers. Educate the loggers to prevent fires. Educate the loggers to come hell-bent for a smoke at the first sniff of it. Educate them in what the making of this black land into a green land again will mean to their children and grandchildren. The loggers can beat The Burn. Loggers can lick anything."

In the early spring, Schroeder organized meetings with The Burn's many loggers in their own areas to talk about fire prevention and fire suppression.

"From now on, we'll hit any fire with everything we have in the beginning, not wait till it gets out of control," he stressed time and again.

He spoke of co-operation, responsibility, and educating the people. He talked the loggers' language, melting their cynicism, and they listened and asked questions. Fire was the great concern. As in a war, it meant the massing of quantities of fire equipment, the construction of new access roads into an intricate network, the building of miles of firebreaks, and the organizing of a standing army of fire fighters during the dry seasons. Schroeder wanted the loggers to be on standby in case of outbreaks. Later, the Northwest District would become unique in the nation, with its year-'round work force alternating among summer fire patrols to snag falling, pest eradication, road construction, and tree planting in the winter and rainy seasons.

Meanwhile, the special forestry committee concluded its many hearings, making its report to Governor Earl Snell. Rehabilitation of The Burn was feasible, the investigators concluded, but it would require many years and much money—from $6,000,000 to $20,000,000. That was a bundle for the state of Oregon in 1946, especially in view of the high risk of The Burn tinderbox. Yet Snell and many others didn't want the federal government moving in. Uncle Sam owned too much of the state as it was. Standing on the firm base of this citizens' study, Snell recommended that the 1947 legislature adopt a 10-year program financed through a $.20-per-1,000-board-feet surtax on all stumpage cut within the state, earmarked particularly for The Burn and also for research and experimental projects in fire prevention, and for the advancement of the forest industry's wood waste research.

The industry, at least part of it, didn't like the surtax idea until it was assured that benefits from the reforesting program would come its way, both over the short and the long hauls. The legislature carried the ball to the people with a constitutional amendment allowing for a healthy bond issue not to exceed at any one time .075 per cent of the state's assessed valuation. The state could thereby raise $10,500,000 for rehabilitation work; the state forester, desiring to play it cool, recommended that the maximum be fixed at $750,000 for any single year.

Oregonians, weary of hearing what might be done about the Big Burn, heartily approved the package by a substantial margin. It opened the way at once for the greatest single forest rehabilitation project ever attempted, one of such magnitude in terms of both size and the potential problems that it even struck fear in the hearts of those who had spoken out favorably in its support. It could be political suicide. Fire, rodents, animals, eroded soil, the weather—all were time bombs against its success. Trying to reforest some 1,400 square miles of rugged wilderness terrain in a decade was difficult, if not impossible, to comprehend. As a young reporter and feature writer, I for one couldn't believe it would ever be accomplished as Schroeder and Kallander outlined it. Yet you needed to remember the wonderful tree-growing qualities of this Pacific Northwest land and place faith in that fact. Foresters had already made a fast start, time being of the essence. Many miles of road and fire corridors were built, the region was being mapped in detail, new lookouts were being raised, and snags were being dropped by the thousands. Bit by bit the state had acquired 250,000 devastated acres and eventually would increase this to over 500,000 acres in northwestern Oregon, the largest of all its state forest areas.

In July 1949 Governor Douglas McKay, who later served as Secretary of the Interior under Dwight Eisenhower, stood on an old stump at the Owl Camp to launch the big project officially. Earl Snell, who helped bring it about, wasn't there, having been killed in office in a plane crash; and Nels Rogers would hardly see it get under way, dying the following October (1949), leaving the Owl Camp to bear the name of that man of vision, along with the memorial forest of the first plantations at this summit of the Coast Range. As the formal ceremonies wore on, power saws were heard down the hill as the plantations were made ready for the first seedlings, and there was the thrash of helicopter blades.

The work went forward slowly, methodically, for a full generation. The Tillamook Burn became a sprawling forest laboratory and gigantic tree farm, and the nation and the world would benefit from what was learned there. The problems of planting the tender seedlings and then protecting them from rodents, deer, bear, rabbits, and the weather, and making them grow to maturity in land that was burned and scarred many times over confounded the foresters, so that on some days there was only discouragement and very little progress.

"At times," said Frank Sargent, "it seemed like we would never get out of the Owl Camp."

Foresters had to feel their way, since there were no guidelines other than Nels Rogers' plan, and he was no longer around to explain it. The State Forestry Department must blaze its own trails. By trial and error, often the experimental efforts didn't work out. One attempt was broadcast seeding by helicopter—the first time the whirlybirds were used for forestry work anywhere in the country. The aerial seeding had a very low per cent of survival—in places virtually non-existent—and there was much waste of seed that was strewn across deadfalls, where it wouldn't germinate or was consumed by rodents and birds. Various poisons were tested against the rodents, which brought an outcry from bird lovers. Finally, with the help of the U. S. Fish and Wildlife Service and outdoor groups, state foresters settled on an overcoating of endrin, which proved effective to protect the seed.

Hand planting, tedious and slow, remained the best method, but there was still much loss of seedlings that couldn't survive the ghastly conditions. Some areas, such as Cook Creek and the East Fork of the Trask River, were so badly burned in 1933 that the soil just wouldn't "take." A quarter century later, foresters were still trying. In the early years plantings were six feet apart, but this proved too close. Later, foresters expanded the spacing to 10 and then 12 feet, since the early settings were overcrowded as the trees began spreading out. It also increased the mileage on seedlings. In some years seedlings and seeds were difficult to obtain in quantity; the demand on forest nurseries was a heavy one indeed. Rural people made extra pocket money harvesting cones in the fall for the state. Planting costs were rising, for delays and setbacks extended the $10,500,000 program beyond the 10-year estimate, continuing for a quarter century before being formally terminated in 1973.

Inflation began hitting the project in the late 1960s and early 1970s. It was fortunate that the program was about finished rather than the state being forced to ask the people for more money, since Oregonians, like most Americans, were by now in a mood to turn down all tax measures by their governments. In 1964 the cost per acre for hand planting was $17.85, but by 1973 it had nearly doubled, to $29.46. Tree planter wages climbed from $12 a day in 1956 to over $24 a day 16 years later. And planters gained other benefits, including sick leave and vacation time. The wage benefits picture also changed in other areas. Time was when college coeds and guys, or couples, willingly spent the summers on lookout duty in isolated ridgetop towers for a set wage. Lookouts of the new age demanded eight-hour shifts, days off, and overtime. Foresters found it cheaper to use patrol planes based at Hillsboro, and closed down many of the lookouts.

Early in the project, perhaps in anticipation of a cost squeeze, state foresters and state prison authorities established a field camp at South Fork for trustees from the penitentiary. The idea worked well, the men being used for planting, road and trail building, and summer protection crews. It paid dividends also in rehabilitation of men, along with rehabilitating the giant forest, for the work in the outdoors brought a strong feeling of long-term accomplishment.

One of the most colorful and effective programs involved thousands of children from public and private schools, church and youth organizations. Each planting season for 24 years, busloads of young people of junior high and high school age made a trek to the wilds to "help replant The Burn," organized by the state, the school districts, and the West Coast Lumbermen's Association. Special plots, some designated as memorial forests, were set aside for the schools, and the experience appeared a lasting one of significance to both parents and children, who could always boast that they had a hand in bringing back this forest. Sometimes the outing moved young people beyond expectations, as in the case of one teenager whose parents were becoming increasingly concerned about his welfare.

"We were very worried about him and the direction he was headed," said the father. "We just couldn't reach him. Then he went on that tree-planting trek and something happened. Maybe it was the thought that those tiny seedlings would be standing long after he had left this earth . . . or that he was doing something really important. Anyway, he completely changed, and we give the credit to that day in The Burn."

If there was one such case, there were likely many more. Some of the teenagers even decided to make forestry their lifetime work.

A full generation grew up with the youth effort. In the last major planting year of 1970, 1,900 students joined the hundreds of thousands who had gone before them by setting out 50,000 trees along the Wilson River. And two years later the James John Grade School of Portland celebrated 25 years of tree planting in The Burn. By then it was second generation, with children sticking seedlings into the ground as their parents had done in the early years. And many mothers and fathers driving along the highways pointed out with pride to their offspring the healthy green timber that "Mom and Dad planted when we were your age."

Foresters learned to pay close attention to the genetics, the climatic conditions, and the elevation of the plantations, so that seedlings were returned to the land of their heritage. Freeze, drought, and scorching sunshine took their tolls. So did the animals. The deer problem became a critical one, for the deer population had exploded in that open country, and the bucks and does thought the young seedlings were set out just for them. The animals feasted on the tender shoots in early spring, when browse was low. The situation became desperate, with 80 per cent of the young trees struck down in 1953–54.

Foresters were confronted by sportsmen whose protective sympathies lay with the deer, since The Burn was one of the West Coast's most outstanding hunting areas. Foresters appealed to the State Game Commission, working out schedules for either sex and doe hunts, and issuing special tags to areas where the pressure on the trees was particularly heavy. In critical sections like the South Fork of the Wilson River, special archery hunts were staged to cut down on the deer population. Little was said about the primary purpose behind such hunts, but the sportsmen were overjoyed at the chances to bag extra deer.

The Douglas fir seedlings themselves also proved tough. Deer would nibble them down to the nubbin, but next year the little trees would come back, a bit taller and stouter, only to be eaten down again. After several years, they were tall and tough enough so that the deer left them alone, but it still slowed the over-all reforestation program, since the permanent loss was so tremendous that at times it seemed reforestation would never be realized. There was also trouble with hordes of rabbits, and when the trees became taller, from black bear, which clawed the bark, killing and maiming the timber. But deer were the worst offenders, and to learn more about the effects of the animals on young trees, a 360-acre area was fenced off into which 36 to 50 deer were run on a rotating basis for an extensive research study by foresters and gamemen.

Many experimental projects were established by both the state and private timber companies. Foresters came to The Burn from all over the world to see firsthand what was going on and to absorb what was being discovered, for the pioneering answers would have a long-range effect on all forestry, not only how to plant trees but in the techniques of protection, road-building, erosion control, and the use of new tools, equipment, and chemicals of modern Space Age technology.

The power saw, particularly that new saw chain model developed by a logger named Joe Cox, proved a boon, since the snag problem was a serious one, averaging 33 to the acre. Teams of three could drop up to 70 a day, although the average was around 40. All those snags had to come out of there, so the power saw arrived just in time, for it would have been a near-to-impossible task with the old crosscuts, powered by human cursedness. Over 1,500,000 snags, 1,000,000,000 board feet, were felled before inflationary wages made the cost prohibitive. Foresters didn't like abandoning the snag falling before it was quite done, for the original plan was to eliminate all those bone-white candles sticking above the ridgetops. But by the 1960s fire protection had improved immeasurably. Also, the new forest was holding the moisture, and the network of 220 miles of fire corridors and 160 miles of easy-access roads made The Burn safer than ever before in its turbulent history.

Schroeder and his aides had worked hard at fire protection, knowing that everything hinged upon it, and flung up many new lookouts for "there are a lotta holes out there," meaning the blind spots of deep canyons and endless ridges. On Saddle Mountain North, foresters built a new lookout by using a helicopter, hauling tools, fittings, building blocks, and lumber topside. It made the job far easier, since the twisting trail was long, narrow, and dangerous for pack animals. In the early fifties, this, too, was a pioneering project, since the versatility of the whirlybird in forestry was just being discovered, from aerial seeding to rescue work.

The foresters lived in constant fear of fire. Three years after serious planting began was another season of the "six-year jinx." Cynical lumberjacks were betting the state foresters wouldn't make it through the summer without a good fire.

"This year tells the story," declared "Molly" Hogan, a grizzled timber faller who'd been dropping stuff for 30 years with ax and misery harp, and was now using one of those newfangled power saws. "The Big Smoke blew up in '33, '39, and '45. It's about time for an extra-dry year. We'd better watch out for the 'six-year jinx' . . ."

Although foresters and data-loaded meteorologists placed no faith in superstitions, there was plenty of talk about a dry cycle in the Pacific Northwest. The thirties was a dry decade, while the forties rated as "a wet one." Now they were into the fifties, perhaps another dry ten years, and there was the added danger of a "jinx attitude." Plenty of rain fell that winter, but as is characteristic in the fir belt, it dries out rapidly when the rains cease. This was the case; even the natives found it difficult to believe after so much deluge. By March 1951 the fireweed and bracken fern lay explosive across the 311,000-acre Burn country, while tree planters still had another month to go. Governor McKay advanced the May 15 fire season more than a month. "Dynamite dry," the loggers said. The air was tense and crackling, with no fog from the coast in two weeks as the last of 3,000,000 seedlings went into the soil to end the winter's planting.

But Ed Schroeder was relying on the loggers and all the spadework that had been done during the past three years. So far it seemed to be working. The 1949 season was a blistering one, with 71 fire outbreaks in The Burn, held in total to 3.11 acres. In 1950 there were only 14 fires, blackening a quarter acre. Yet The Burn could still blow up again, and this might be the time.

On April 23 sparks of a power saw ignited the parched ground cover of an Elkhorn logging show. Flames spread to the blowtorch snags, traveling fast from top to top. The alarm sounded; from all directions loggers converged as never before on the scene, fighting fire side by side under the orders of the foresters and state suppression crews. It was an unbelievable sight, that would have pitched old-time jacks into the nearest saloon. The fire was contained within 7,400 acres.

Yet all spring and summer the fire smoldered, closely patrolled by men and equipment against any further outbreak. Only the fall rains would permanently wipe it out. But this was a worsening fire year along the West Coast. Thousands of acres were being ravaged, from Southern California to British Columbia. The Burn lay dormant like a smoldering volcano until the afternoon of July 11, when it exploded along the North Fork of the Trask, raging through snags, fireweed, and vine maple, scaling the peaks and ridges in country that would be thought impossible to burn again. And once more, a familiar mushroom cloud gave out an eerie blood-red glow in the sunset, for this was a hot one, which threatened to take off and was menacing the village of Trask, building its own updrafts and hurtling great snags for 100 yards across the canyon.

Schroeder's troops hit the fire hard with 23 bulldozers, 60 power saw teams, 18 tank trucks, a dozen portable high-pressure pumps, and all of it effectively co-ordinated by two-way walkie-talkie radio from a high ridge observation point looking down on the flames. Again, the loggers were alongside the foresters, and the fire was contained at 2,800 acres for the rest of the year, the first time that two major outbreaks in one season had been controlled without running wild, and this time possibly destroying valuable first plantings of the new forest. The "six-year jinx" was broken; Schroeder and his team knew it; and Alex Scott, a veteran logger, who had a half section of ranch at The Burn's edge and who had fought all the fires, making a "last stand" in 1945 with his son at the outer fringe of his forest farm, summarized it:

"We've made a real start at last to bring The Burn back to be a forest—good green country like my half section that was burned in 1904."

Chapter XIII

Wood Worm Power Saws

The object of my invention is to provide mechanical means for sawing down trees which may be easily and quickly manipulated by two men with such rapidity that the largest and toughest trees may be felled in an exceedingly short length of time.

> Patent application
> About 1900
> From *Southern Lumberman*

The biggest single task in rehabilitating the Tillamook Burn, next to the tremendous job of replanting all that mountainous country, was ridding the sprawling devastation of armies of unsightly, fire-dangerous snags, which appeared like porcupine quills across the landscape as far as the eye could see.

Under the traditional fell-and-buck methods, with crosscuts powered by loggers' sinew, such an operation would take a decade or two and be exceedingly expensive. But now there was a new sound in the brush, a screeching and screaming yet unfamiliar to many loggers and lumbermen. The chain power saw was coming of age.

As had been the case with wartime salvage work in The Burn, World War II provided the impetus for power saw development, some of it stolen from Nazi Germany. But the trump card was once again played by a lumberjack with that native ingenuity and inventiveness that led him to follow in the footsteps of John Dolbeer, Ephraim Shay, and all the unknowns who through the decades of logging and lumbering had somehow found new and better ways of applying the nation's mechanization to their own peculiar trade. His name was Joe Cox, his interest was wood worms, and his is an American success story that ranks with the best.

The unhappy chore of cutting down trees and slicing them into manageable logs hadn't changed much since the first settler of the Atlantic seaboard dropped the first tree for the first log cabin. The cursed misery whips were Custer's Last Stand against the Indians, the lone holdout of an age that was almost gone. It wasn't simply a reluctance on the part of fellers and buckers to modernize, although that was part of it, for loggers didn't appreciate the smell of petroleum and carbon monoxide. It was, first and foremost, that many early power rigs were not only cumbersome, but also simply didn't do the job. In a way, it was strange, for mechanization had come to the woods as it had to the lumbering end of the game, and power rigs were being constantly changed and improved. The advent of the gasoline engine and then the diesel caused another "revolution" in getting sticks to the mills, so that steam power was bulldozed out of the woods. Logging outfits were ripping up their railroads and junking their proud locomotives in favor of the internal-combustion engine, turning railroad beds into roads for wheezing trucks and even replacing steam donkeys with gas engines to yard the sticks and load them on the trucks or railroad cars.

The changeover to the gasoline-driven power was the beginning of the final chapter of the timber industry's great industrialization, a revolution that is still going on today. Yet while all else—save perhaps for the poor choker setter—was being mechanized, the felling and bucking of the trees remained stagnant. Lumberjacks still stood on springboards and leaned into the great crosscuts, even though they were beginning to live in the towns with wives, families, and home comforts, commuting each day in their own private cars to the woods.

There had been many attempts to develop some kind of power saw, but mostly these were unwieldy affairs, difficult to handle and move from tree to tree, and often undependable, breaking down under the rugged demands of dropping these hefty giants. Visions of a practical, portable power saw extended back generations, and the dreams were vivid among the older fallers who were beginning to feel their years as they fought the thick hides and inner fiber of the big trees, and came home with aching backs. The power saw had been considered for at least a century, but just what form such a rig should take was a matter of debate and rough opinion among loggers, dreamers, and would-be inventors. There was a definite leaning toward some kind of "chain" saw that would link saw teeth into an endless belt to be powered through the saw kerf to make the cut, rather than the style of rig where the motor would drag a saw blade across the wood in the manner of the timber fallers.

In 1858, the first chain saw patent was issued by the U. S. Patent Office to Harvey Brown of New York City for an "endless sectional sawing mechanism" of insert saw teeth. In reality, it was a band saw with a series of hinged sections riding on hexagonal pulleys and operating in a slotted guide plate. Five years later, a second patent was issued to George Kammerl, also of New York, under the title "Improvement of Endless Saws." Kammerl described his invention as "a saw blade that is formed of an endless chain, the outside edge of which carries the required teeth adapted to cutting wood or other material, this whole chain or chain saw blade moving continuously in one single direction over a system of grooved pulleys, one or more of which receive the required motive power. . . ." Unlike its predecessor, the Kammerl rig had no guide plate, relying instead upon pulley tension, and this arrangement was concluded to be fundamentally unsound.

Apparently neither of these ideas got off the drawing board, although the basic principle of today's power saws was very evident. The vision of drawing some kind of linked teeth through the saw kerf was there from the beginning, and certainly others had the idea before 1858, for it seemed the natural thing to anyone bent on the agonizing task of sawing wood. The chain idea may have been conceived abroad. But in the latter half of the nineteenth century, at least a dozen patents were issued in the United States, although none of them considered the matter of power nor were clear as to whether the saw would be stationary or portable. And no chain saw was manufactured. Ironically, too, the sawmills had long since found methods to power circular saws for cutting lumber, through water and then steam and electrical energy, where once they, too, had been hand and muscle operations. But the system used by the mills wouldn't work in the brush—at least not for a long time.

All the while, there were constant efforts to develop other kinds of power saws, many of them extensions of the sawmill style. An old print, first published in 1871, showed operation of a steam-powered portable rig employing a saw blade similar in shape to the regulation misery harp, which was attached to a power wheel drawing it back and forth through the kerf, with steam power furnished from a portable boiler. The "Ransome Steam Tree Feller and Log Cross-Cutting Machine" was shown in the sketch being used for both felling and bucking. The efficiency of the drag saw had long been accepted, but this huge and obviously cumbersome rig apparently didn't fill the bill, although one group of visionaries maintained that the nucleus was there; that this was the obvious route to follow in developing a workable portable power saw for the woods. Others contended that a full break should be made with the saw, that some new kind of cutting tool needed to be developed, such as a high-speed cutter head or boring machine. A third group favored the chain as the way to eliminate this last hand job in the woods. As for the power, it could come from steam, gasoline, or maybe even air.

In England, a hand-operated chain saw was patented in 1900 by a William Stanley of South Norwood. But the real thrust of chain power saw development came in the Big Woods of the Pacific Northwest and in the northern California redwood empire, where dreams were replaced with stark realities, as in the case of the mythical loggers' skyhook, which would one day become the reality of balloon and helicopter logging.

Communications weren't what they are today, so that new ways of doing things could take a long while spreading from camp to camp, perhaps first carried by the transient loggers. The exchange of ideas was one of the prime purposes of the Pacific Logging Congress and similar gatherings of loggers and lumbermen. In the summer of the Lewis and Clark Fair in Portland, an experiment was going unnoticed down in the redwood forests of northern California. At Eureka, near Sequoia Park, the first chain power saw operating from a gasoline engine on the West Coast, or perhaps the country, was being tested. The name of the inventor is lost, but a number of lumbermen witnessed the event. The chain saw rig, a strange-appearing contraption, was powered by a two-cylinder water-cooled marine-type motor, with fuel and water coolant provided from tanks suspended on a tree high above the engine level. And it worked; the saw cut through a 10-foot log in 4.5 minutes, which in itself was something of an achievement.

The following year (1906), the Potlach Lumber Company of Idaho developed a crude log deck chain saw operation, which again was strictly a local contraption, although it was noted by Charles Wolf, who improved upon the idea in 1908 for the Blackwell Lumber Company of Couer d'Alene, Idaho. Still nobody took the trouble to patent these machines and promote the equipment for widespread sale, although the Blackwell saw was later manufactured in limited quantity by the Union Iron Works. The problem remained the terrible size and weight of the power equipment.

Others got into the act; any number of hard-bitten practical lumberjacks tried their hands at developing a workable chain saw, alongside logging machinists, mechanical engineers, and backwoods inventors, of which the industry spawned so many. Charlie Wolf wasn't finished by any means. He spent a decade studying the chain saw problem from every angle, and then in 1920, under his own patents, he introduced what ranks as the first successful portable chain saw, putting it into production at his own Peninsula Iron Works in Portland, Oregon. He called it the "Wolf Electric Drive Link Saw," powered by an electric motor driven off a portable generator and amazingly manufactured in three sizes, which weighed from 70 to 90 pounds.

The power saw complication was a dual one—the development of a small yet powerful motor, and the manufacture of a stout and effective cutting chain. The Wolf equipment dealt with both basic factors in an effective manner, at least for the time and place. The saw chain was similar to the conventional two-cutter-tooth crosscut, with its cutters and rakers on separate links that could be easily replaced, and also could be filed and set in the same manner as a standard crosscut. The Wolf saw operated in either direction, a distinct advantage, for when cutters and rakers were dulled from moving in one direction, the drive could be reversed, bringing the opposite cutters into action. Another innovation was that the cutters and rakers played in opposition to each other, thus offsetting the bucking tendency that plagued other pioneer chain saws. This improvement also made it possible to employ shorter chain lengths and smaller sprockets, thus eliminating the need for tension, which resulted in excessive wear.

The Wolf saw was certainly a huge step forward, and with a few modifications, it might have put the industry of the twenties well in advance of the power saw revolution that eventually came about. But the Wolf saw was too far ahead of its time, for suddenly the manufacturer was square against an unanticipated problem. The loggers balked, setting their calked boots solidly against using these mechanized gadgets. What had been good enough for their fathers and grandfathers was good enough for them! As they were doing with the early foresters, the loggers sneered at the power saws and ran the salesmen from the woods. There was also the age-old fear that the power saws would eliminate jobs. And any failure of the gear, such as a lost nut or bolt, a power failure, or a general breakdown, was seized upon by the timber beasts as ample demonstration that the new rigs were nothing more than nuisances.

The adage that "a prophet is not without honor, save in his own country" held special meaning for Charlie Wolf. Ironically, with this vast timber industry right at his doorstep, for which he built and was manufacturing this saw, Wolf was forced to turn to other markets—the construction trade, general contractors, the military, railroads, pole yards, mines, and shipbuilders. Between the two world wars, the Wolf saw enjoyed virtually an international monopoly, although there were potential competitors in the States and abroad who were attempting to

41. Before much planting could be done, the land of the Burn had to be freed of snags and debris, making it safe from future fires. This was one of the first of thousands of snags to be felled. The power saw was developed just in time. —*Ellis Lucia photo*. (Chapter XII)

42. An early planting crew sets out tiny seedlings by hand along a slope of the Burn. The task seemed tedious and slow, but hand planting proved the best method of doing the job. —*Author's collection*. (Chapter XII)

43. Thousands of Oregon school-children made annual treks to the Tillamook Burn to set out trees. The program was one of the most popular phases of the rehabilitation project. —*Author's collection.* (Chapter XII)

44. Rehabilitating the Burn became a giant experimental project in forestry, with many new techniques and ideas attempted. The helicopter was used in forestry for the first time for aerial seeding and rodent eradication, photography, and even for construction of lookouts. Note the hopper for seed on side of this antique whirlybird. —*Ellis Lucia photo.* (Chapter XII)

45. Albert K. Wiesendanger made a second career as head of the Keep Oregon Green program, and was a pace-setter for shaping the national effort to reduce fire in the woods caused by public apathy and carelessness. —*Author's collection.* (Chapter XI)

46, 47. Licking fire in the woods, through better protection, stronger laws, and public education, encouraged investment in new forests. Some three decades ago this hillside above a small logging camp looked like this. The second view shows how it appears today. —*Weyerhaeuser Company.* (Chapter XI)

48. The Big Woods was changing again, as the internal-combustion engine and trucks began replacing steam and the iron horse. Logging operators ripped up their rails to make way for the trucks. —*Oregon Historical Society.*

49. Last holdout against mechanization in the woods was at the far end, in falling and bucking timber, where musclepower prevailed until after World War II. Early power saws were cumbersome, like this heavy Wade Saw, vintage about 1930. —*Oregon Historical Society.* (Chapter XIII)

50. Logger Joe Cox got the idea for a unique kind of saw chain from watching woodworms at work. —*Omark Industries*. (Chapter XIII)

51. Noticing how the woodworms cut to the side rather than scraping, as did early saw chains, Joe Cox designed his remarkable saw chain to do likewise. This closeup shows how upper barbs cut and chip left and right to the side. —*Omark Industries*. (Chapter XIII)

52. When snag falling began about 1948 in the Tillamook Burn, forestry crews turned to power chain saws, which were still heavy and required two men to handle them. Smaller, lighter-weight saws and Cox's chain were on the way. —*Ellis Lucia photo*. (Chapter XIII)

53. Today's chain power saws can be handled by a lone logger, so that at long last the tree falling and bucking has been fully mechanized. —*Omark Industries.* (Chapter XIII)

54. Isolated deep in Oregon's Coast Range mountains, Valsetz is a company town that remains today little changed from the typical logging community of yesteryear. *—Ellis Lucia photo*. (Chapter XIV)

55. Dorothy Ann Hobson as a young girl edited the *Valsetz Star,* which gained her a national reputation. *—Courtesy Jack Pement, Oregon Journal.* (Chapter XIV)

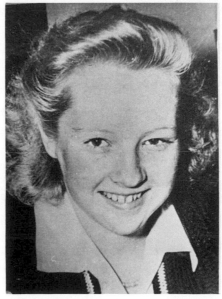

56. Longview, Washington, "the planned city," was built by a visionary lumberman in the 1920s. Unlike any other logging-lumbering town, Longview became noted for its wide boulevards and beautiful parks and streets. The beauty is still there and so is the wood-manufacturing industry. —*Jan Fardell photo.* (Chapter XIV)

enter the business, believing that eventually the power saw must be accepted in the woods.

Much of this latter-day pioneering took place in Europe. Smaller gasoline engines were being developed successfully, especially in Germany, where Andreas Stihl was finding ways to reduce the cumbersome size of stationary deck and pond chain saws for loggers' use in the Black Forest. In 1925 Stihl managed to develop a two-man chain power saw weighing 140 pounds, which was nevertheless "portable." His first one sold quickly after a demonstration. He built another and another, and at the Leipzig Fair of 1929, two saws brought eager orders from neighboring countries, opening the way for Stihl's company to develop a huge export business, which turned it into one of the largest chain saw manufacturers in the world. Yet as in the United States, his salesmen were confronted by the same kind of opposition that Wolf met; the loggers chased them off and wouldn't consider anything to replace the misery whips. Germany was in economic straits, and the loggers feared for their jobs.

The Europeans' small gasoline engine was a threat to the Wolf saw. To keep pace, Wolf offered the saw with either a pneumatic motor or gasoline engine as well as the electric unit. Beyond this, most American manufacturers were dragging far behind, for they were still hung up on the matter of weight, such as the Dow Pump and Diesel Engine Company of Alameda, California, which in about 1933 made a gasoline-driven saw mounted on bicycle wheels. The saw weighed 490 pounds, much too heavy for easy maneuvering, and was also unwieldy for shifting from location to location in the forests, and therefore was a magnificent failure. There were more attempts to develop an acceptable machine, even a throwback to the original Kammerl idea, but nothing worked, and there actually seemed to be both a lack of demand and enthusiasm, which left little room for much progress, since a dire necessity and a mounting demand are what spark American ingenuity. Meanwhile, as lumberjacks continued pulling the primitive crosscuts, the woods were making amazing forward progress in getting out the logs, in fire fighting, fire prevention, and even in living conditions. But if the lumberjack cared about power saws, he didn't exhibit much interest.

Then in 1936, the year of the great Bandon fire, the H. A. Stihl Company of Germany began exporting a unique 46-pound chain saw with an excellent lightweight engine, which immediately caught the eye of the Depression-ridden timber industry. The saw was exported first to the pulpwood areas of eastern Canada, to be purchased and tested by a cautious few. It drifted West to the timber lands of British Columbia, and by the following year, an ambitious agent was exporting them to a Seattle mill and mining supply company. A few were sold, but there was still that hanging back.

However, at the same time, the U. S. Forest Service launched an "equipment development laboratory" in Portland under the direction of T. P. Flynn. One of the laboratory's first projects was to test, rebuild, and modify chain saws for application in the Big Woods. Demonstrations were being held at the logging congresses and wherever else large numbers of lumberjacks and lumbermen were likely to congregate. The Stihl saw was now being vigorously promoted. The upshot of all this, along with the investigation by the forest laboratory, spurred interest in the new saws, at long last improving the climate for acceptance in the woods.

But just as things were beginning to take hold, Hitler marched into Poland

and the Stihl supply was cut off, perhaps forever, from the American market. When Pearl Harbor brought a crisis on the home front, loggers left the woods for higher-paying wages in the shipyards. Lumber became a critical war material, and as in World War I, the military demand was mounting just as the woods felt the pinch of the labor shortage. Suddenly, power saws were in great demand by the timbermen, and at the same time, the Army and Navy were also calling for them.

Stihl was completely out of the picture, being on the side of the enemy, so that D. J. Smith and the Mill and Mine Supply Company of Seattle, which were marketing the saws, had to begin manufacturing them to stay in business. In 1938–39 there was only the Wolf saw being made on the North American continent, by the Reed-Prentice Corporation of Worcester, Massachusetts, under the Wolf patent and an agreement with the Portland firm, which couldn't meet the economic production problems. But by Pearl Harbor, there were a half-dozen outfits in power saw production, mostly gas-driven and patterned after the Stihl rigs, which no longer enjoyed patent protection. Americans were not only interested in the power unit, but also copied the Stihl chain, which had several advantages: It could be manufactured cheaply, and was reasonably tough and rugged; the logging industry was somewhat familiar with it, making it most readily acceptable; and in character with the American business mind's tendency to cut costs wherever possible, the Stihl chain was up for grabs, with no royalty payment required. Thus, after almost a century, the power saw became light enough and dependable enough, especially under wartime pressures, to be acceptable to the American lumberjack.

Power saws needed to be tough to withstand the rigors of the Northwest forests and those huge trees. Gradually, improvements were made in sprockets and bars, the latter needing a tough, light metal to keep them in true line during the cutting process. But there was also dissatisfaction with the cutting chain, which clung to the traditional approach of the old crosscut's teeth being dragged through the saw kerf, tearing and ripping at the wood fibers, or the principle of running a sharp nail over the surface of a piece of wood. The principle worked, but the cutting action was very rough, and there was a lot of waste. Also, the saw chains didn't hold up too well, and as much time was lost in resharpening the cutting chains as was spent in actually felling the big stuff. Some concluded that the old crosscuts were more efficient, and returned to them.

Along came Joe Cox, a thin-faced logger with a sharp jaw who had knocked about the West for a number of years and finally wandered into the forests of the Pacific Northwest. Cox, who was born in Oklahoma, wasn't strictly a lumberjack; he'd left home at age 16 for Colorado to work in the shops of the Rio Grande railroad at Grand Junction, where he stayed two years. By the time he came to the Northwest, he had served a solid apprenticeship, self-made, as a machinist. Following the railroad stint, he and his brother-in-law began an automobile agency at Fruita, Colorado, and then a bus line. In those enterprises, he learned a lot about mechanics, motor repair, and welding, which landed him a job building a new gas line to San Francisco. Then he worked on the construction of powdered milk processing plants for the Golden States Milk Company, an intricate and complicated project. Cox didn't have a degree, but he was an engineer in every sense, typical of his generation, which learned the tough way how to make things work through hard-bitten trial and error; and he had an unbridled ingenuity and willingness to attempt anything that might provide an answer to a mechanical problem.

Cox moved to Arizona, where he set up a small welding shop with a brother. They repaired cars and wired many homes. He also designed and manufactured a highly successful water heater for home use. When things slowed down, Cox shifted to Texas, where he worked at Hobbs, welding oil-drilling equipment. Then the brothers decided to try Oregon and were suddenly confronted by a new world of logging and lumbering.

"We felled, limbed, and bucked small frozen knotty pine timber in three feet of snow our first winter here," Cox remembered. "We were paid $.50 a thousand. We earned about $4.00 for 10 hours of hard work, and it was hard."

Cox "went through the chairs" in the woods, learning every job. He couldn't help noticing that there was a decided movement toward mechanization in logging the stuff, which sent the inventive wheels spinning in his own mind. One morning near Chiloquin, east of the Cascades, the brothers were confronted by a strange contraption, which they were asked to try out. It was a power stump saw, driven by a motorcycle motor and mounted on wheels like a pushcart. The power saw used a laminated bar about a foot wide, but was quite inefficient. The rig just didn't cut very well.

"We couldn't fall a tree as quick as we could with a hand saw," Cox recalled. "This seemed strange to me because the power saw had plenty of stuff. I was a pretty fair filer at the time and figured that if I could make a power saw cut as efficiently as a crosscut, it should practically fall through the wood. It just made sense, and with such a cutting tool, sawing timber would be a lot easier."

Filing the notion in his mind, Joe began trying to come up with something that would do a better job than the scratcher chains. Somehow he felt the answer would be found in Nature herself. One day while cutting firewood, he laid his ax into an old stump. Suddenly he stopped, for he'd cut into a small tunnel of nature's most celebrated and destructive woodsawyer, the timber beetle, or scientifically, *Ergates spiculatus*. The larvae of this beetle, cursed in the kind of verbiage formerly applied to oxen by the old bullwhackers, have an amazing ability for cutting and destroying huge quantities of timber, although the busy grub is hardly the size of a stout man's finger. The hated grub turns good timber into sawdust, and it doesn't matter whether the trees are alive or sound snags and windfalls that might be salvaged. The vast Tillamook Burn and other disasters promised feasts that would last the timber beetle and his offspring many generations, although the way they worked, wholesale destruction of salvageable timber acreage might be accomplished in a few brief years. The winged adult beetle deposits its eggs beneath the bark of a dead tree, or when faced with an overpopulation problem, under the bark of living trees. The results are disastrous.

Logger Cox, hoping none of his lumberjack friends were watching, suddenly was asking himself, "How do they do it?" as he bent over the tiny tunnel of that hollow log. Although he was readily familiar with the cursed beetle, he'd never thought about them before in the terms now coming to his mind. What amazed Joe was the force with which the grub cut through sound timber, either with or across the grain, at its own will, and leaving an astounding pile of sawdust in its wake. He discovered that the grub used its head, which was armed with two sharp, tough woodcutters. The lumberjack, securing a magnifying glass, spent many hours studying the action of *Ergates spiculatus*. For long stretches, he watched the grub in action, considering its daily habits and its method of cutting. He put the sawdust left in its wake under a microscope. He was most intrigued by the grub's cutting action, left and right, side to side, rather than scratching or burrowing directly ahead. It was an amazing technique that started to haunt Joe Cox. Something began germinating in his mind.

In the basement of his two-story frame house in southeast Portland, Cox designed a different kind of cutting chain, applying the principle learned from the timber beetle, of cutting left and right and to the sides, rather than scratching away at the wood. The result was a revolutionary chain that not only cut faster and better, but also could be used longer without resharpening.

Cox patented his idea, but it was some time before he could get his chain into the market. He worked as a blacksmith and sold saws as a field man to make a living. In 1947 he established his own company, called Oregon Saw Chain, and continued his testing on various kinds of timber, including those two giant sequoias felled amid headlines at Forest Grove. His small manufacturing "plant" was the basement of his house, where he set up two assembly tables and had a payroll of four. The first year, the firm sold $300,000 in saw chain, for the Cox product apparently really worked and thus cleared the last remaining hurdle for use of these power rigs efficiently in the woods. The lumberjack couldn't keep up with the demand in his home. The following year, he moved the operation to north Portland, into an old converted repair garage. The payroll jumped to 12 persons, one of them a personable 28-year-old graduate of the Harvard School of Business as assistant general manager. John Gray was given a keg of nails to sit on at a big work table, which he shared with the general manager. And Joe Cox was running around, supervising everything in coveralls, sweatshirt, and baseball cap, and projecting his ideas for the machinery that was turning out this unique saw chain.

Cox's jackpot idea forced them to move again two years later into a building that could accommodate some 70 workers, and by 1955, with sales reaching $7,000,000 into a permanent location on the southeast edge of Portland, where the payroll grew to 2,800 employees. The company continued growing, for the demand for Cox's bug chain became worldwide. By its twentieth year, sales reached $49,000,000, roughly 163 times what they were in the first year and with plants in many foreign lands. This entry into the foreign market caused a name change, on urging from the sales force, to "Omark," derived from the phrase "Oregon mark," using the letter "O" and the latter half of "trademark." "Omark" was far easier to pronounce and understand in lands abroad. But the chain itself continued to carry its original brand, Oregon Saw Chain, and although the firm branched out into the manufacture of allied equipment, it has left the chain saw power units to others.

The saw chain manufactured in some 35 sizes by Omark would doubtlessly stretch around the world a good many times in a quarter century. The multimillion-dollar company has expanded from chains and bars into other fields, including masonry and concrete-cutting equipment, sports arms and ammunition, and fastening tools for the aerospace industry. The power saw today has worldwide acceptance. There are photographs of Africans and South American Indians, in loinclothes, using the chain saws deep in their own heavy forest jungles. They're more than logging equipment now, owned by thousands of private citizens, who consider them a necessary tool for homes and private property, such as a second place in the mountains or at the coast, for keeping down the brush, clearing the land, or cutting winter's wood from the driftwood logs cast upon the beaches, and for camping or fishing trips. Most of them use Joe Cox's bug chain. And the old weight problem has been licked, too. The McCullough people, for one, have a power unit that weighs slightly less than seven pounds and will cut through a five-inch log in six seconds. You can't do much better than that.

As for Joe Cox, he became wealthy, went on to other inventive glories, turned things over to John Gray, and sought a warmer climate at Santa Barbara, California, where I wonder if he ever thinks about that fateful day when he swung his logger's ax into that hollow stump and his eye lit on the tunnel of *Ergates spiculatus.*

Chapter XIV

Company Towns and the
Valsetz Star

If we had a war, Hitler couldn't get over the hill to get us because he would get stuck in the mud.

DOROTHY ANN HOBSON
Valsetz Star

Wauna was essentially a "dead" area when Crown Zellerbach decided to locate a big new paper plant on the lower Columbia River in the mid-1960s. The staggering blows of the Tillamook and Wolf Creek fires, followed by the wartime demand for timber, had all but eliminated the fat veins of green gold from the northwestern Oregon region. Like a mining camp of the Old West, when the trees petered out, Wauna became almost a ghost town.

Now the boom times were returning. The impact of 1,400 construction workers followed by a permanent population of some 650 people sharing a $6,500,000 payroll would alter the downriver settlement drastically. But other than houses for its managers, Crown Zellerbach didn't build a company town at Wauna, as it might have done in timber's raw age. The employees were encouraged to scatter up and down the river to avoid heavy impact upon local schools, housing, water supplies, and sewage disposal systems of existing small towns in the Wauna area. Life for makers of wood products had become more like that of other workers. Since modern timber plants are too huge and costly for overnight moving, like a circus, the Wauna plant would likely be around for a long while.

In the untamed years, the mills and the villages or camps followed the timber supply and, as we have noted, the logging and sawmill operators built "company towns," owned and governed by the outfits, which provided workers and perhaps their families with all life's necessities—houses, stores, recreation halls, churches, and even schools. There was also the benefit of having much of the payroll plowed directly back into the owner's treasury in the form of "rent" and through the company store. And while the timber baron might feel he was being very generous in furnishing convenient facilities, the rental fees and prices at the only store within reach were often as high as the traffic would bear. There was little the isolated residents could do about it; to protest meant the loss of your job, since there wasn't any citizens' government, and company edict was the law of the land.

But like all else in the timber industry since World War II, the life-style of the logger and sawmill worker changed with the widespread use of gasoline and diesel-driven power. As the power saw saved the backs of fellers and buckers, the private automobile and better roads took the lumberjack out of isolation, to live where he chose and commute to his job just like a city feller. It also freed him from the 'round-the-clock rule of the outfit that paid his wages. The isolated logging camp and company town began to fade until, today, they have largely vanished. Few remain outside the wilds of Alaska and British Columbia; there simply isn't that much isolation. Yet the Big Woods still harbor their own unique timber towns, each one distinctive from all the others.

Grisdale, on the Olympic Peninsula, is perhaps the outstanding example of the new-type company village, a far cry from the lice-ridden shanties flung up in the wilds fifty years ago. Pope & Talbot still operates its picturesque company town of Port Gamble, more than a century old and neatly unique in appearance and concept, as generally it has always been. Port Gamble has been refurbished in its New England best, for it is now a National Historic Site. Then there is beautiful Longview, Washington, celebrating its Golden Anniversary in 1973, and probably the slickest timber city ever designed and built by a lumberman. As for the traditional company logging and lumber town, I know of only one: Valsetz, deep in Oregon's Coast Range and, in appearance at least, a rustic throwback to the roaring times of yesteryear.

The lumbermen took a vital interest in the towns, whether or not they owned them, and one immediately thinks of tiny Gardiner on the Oregon coast, which for many decades could be easily spotted as a company village because all the houses and other buildings were painted sparkling white, with blue roofs. The Simpson people, primarily through the efforts of Mark Reed, contributed much to the development of Shelton, Washington, as a pleasant place to live, for Reed exhibited a feeling of responsibility for the community, which flourished because of the company. Other timbermen held ambitions and dreams to build and improve their towns, but none reached the epitome of R. A. Long and his planned city.

Following World War I, the Long-Bell Timber Company found its resources in the South depleted, and invaded the Pacific Northwest. There was little choice, for otherwise the company would certainly die. Robert Alexander Long, a conservative and pious man who read his Bible and offered prayers to open his board meetings, came out to look the country over. In Cowlitz County, Washington, a lush forest caught his eye, so he made a deal with the Weyerhaeuser people for 70,000 acres, containing several billion board feet. As for a mill site, Long and his associates decided upon the soggy lowlands where the Cowlitz River dumps into the Columbia—a historic spot known as Monticello, where in 1852 Washingtonians gathered to petition Congress to establish the Washington Territory, separate from Oregon.

Long wanted the very best; no haphazard village of clapboard buildings for him. He employed J. C. Nichols of Kansas City, a landscape architect no less, to help plan his city. That was a distinct departure from tradition, for no Northwest lumber king had ever considered such a thing. In the summer of 1921, work started in earnest as crews began diking the lowlands and installing an intricate drainage system, while surveying was going ahead, with streets and lots being platted.

R. A. Long may have been pious, but he wasn't necessarily modest, desiring that the town be named for him. That was fine enough, but it proved a sticky matter with the Post Office Department, since there were not only nine Longviews in the United States, but one was unfortunately located in another Washington county along the banks of that very same Columbia River. The mail authorities wouldn't allow two towns of the same name within the same state. The lumber people would have to come up with something else, perhaps Monticello.

Impossible. Something must be done to eliminate the other Longview. Long-Bell officials, with a stroke of master salesmanship, convinced the established village to change its name to Barger. An agreeable local postmaster signed the petition. But as company envoys were about to leave, a citizen remarked that a tiny depot would be most welcome to cover the railroad platform, for the mailbags tossed from passing trains often landed on wet and soggy ground. The Long-Bell men were happy to comply by erecting such a shelter, which cost the terrible sum of $25.

"The only thing cheap about Longview was its name," wryly observed John M. McClelland, Jr., the city's historian.

Other costs soared into the wild blue yonder. The 14,000-acre townsite was $125 more per acre than had been estimated; and Long-Bell engineers and officials were way off in money projections. The diking system, which was estimated at $817,359, actually cost $3,250,000. Sidewalks, parks, lakes, and the broad avenues were far above calculations. The company also had to build a railroad, since the Great Northern, Union Pacific, and Northern Pacific didn't want Long-Bell locomotives and cars moving on their pooled trackage. Long-Bell formed a subsidiary company designed to frighten the big outfits with its name —the Longview, Portland, & Northern. It never got far beyond Longview, save for nine miles up the Cowlitz to the company logging camps, but the railroad cost a pretty penny, $5,400,000. Added to all this, two large and costly sawmills had to be built.

The "Model City of the Twentieth Century," laid out ambitiously for over 50,000 people, had along its broad boulevards a six-story hotel appropriately named the Monticello, a brick colonial-styled city hall with tower and clock, several schools, railroad station, library, and numerous business buildings grouped around a landscaped common quite naturally called R. A. Long Park. Long financed many of the public buildings from his own personal fortune. Far ahead of his time, Long paid much attention to the new city's environment, even to a huge man-made lake to be surrounded by lovely green trees, lawns, and flowers, and named in memory of Lewis and Clark's first lady, Sacajawea.

At the end of July 1924, thousands gathered for the formal dedication of Long's dream city with a Pageant of Progress. The "biggest sawmill in the world" cut its first logs, while a band, with R. A. Long's persuasion, struck up "Nearer My God to Thee." There was an endless stream of tireless speeches praising the founder by the governor of Washington, a United States senator, a congressman, and lesser dignitaries, including lumbermen who streamed there from far places to salute Long for his achievement and vision of progress, which would help the image of them all. On the final day, the eminent evangelist Billy Sunday, secured by Long for a cost of $500, gave a rousing sermon, declaring that he "wouldn't preach outdoors for anybody but God and R. A. Long." The only sour note came from the IWW newspaper, which dutifully reported the event under a classic headline: "Long-Bell Hires Jesus Man to Quiet Slaves."

Nevertheless, Longview survived as a thriving timber town and seaport, and today ranks as one of the Northwest's most beautiful cities. But the flavor of the forests and the industry is still strong there, where logging is taught for credit in the local R. A. Long High School.

Valsetz, the oldest company logging-lumbering town in Oregon, is 15 miles into the back country over a twisting gravel road, and 50 years back in time. Valsetz derives its name from "valley" and the nearby Siletz River, and hasn't changed much since Cobbs & Mitchell established a sawmill there many decades ago and built the town for its workers. It's also much as Dorothy Ann Hobson, girl editor extraordinaire of the dark days of the Great Depression, described it to her international readership of the celebrated *Valsetz Star,* with unpaved streets, single long concrete walkway, clusters of bunkhouses, unpainted frame homes, repair shops, well-worn grade and high school buildings, a dormitory for 15 teachers, volunteer fire department with single seldom-used engine, the dust of summer, and the mud of winter in deluge that runs upward to 160 inches annually, making Valsetz the wettest place in the Pacific Northwest, if not the United States.

Some 500 people live there, for Boise Cascade operates a combination lumber and plywood mill and owns the town. But company spokesmen prefer not to talk about Valsetz, rather to dwell upon Boise Cascade's achievements in sleek industrial plants, pollution controls, and balloon logging. Social life centers around the company store, adjoining cafe and recreation hall, where burns the only lighted sign in town. Otherwise, Valsetz appears to be a semighost town that once had a proud community band, trophy-winning teams, theater, pool tables and bowling alleys (still in use), and the great outdoors to relieve the depression from months of gloom and downpour and the bitter cold of winter. Thanks to cable television, the outside is much closer now. The road is improved, too, while the old railroad still hauls out the wood products; and its citizens are just as opinionated over the state of the nation and the world as when Dorothy Hobson ran the *Star*.

"One night I just got mad, yanked the cord out, and turned the TV facing the wall," said Wayne Hadley, a 30-year resident. "We sold it later. Too much junk on it when our kids were little—illicit love affairs constantly, always violence. Children get the idea you live by violence."

Valsetz citizens have always had to depend largely on their own resources.

"The winter of 1929–30 was long and bitter," remembered Mrs. B. M. (Crystal) Rose, who spent six years there and was happy to leave. "It was so cold that my baby had her hands frost-bitten when she got them out of the bed covers one night. We were snowed in for several weeks. The train couldn't get in, and we couldn't use the road. Food was scarce at the company store, so we shared. If I had a bit of meat and the neighbor had potatoes, we ate stew. Pipes froze, and we melted snow for water. We thought we had it rough, but when the real Depression hit us, we found out the meaning of privation. . . ."

Valsetz was first called Sugarloaf, for the nearby mountain. It had its humor in streets and districts named Snooseville, where the Swedes lived; Shanghai, Cadillac Avenue, and Snob Hill, where foremen and supervisors had their fancy places. But nothing quite licked the isolation. The safest way in and out was via the Valley & Siletz Railroad, for the automobile road was so narrow that two cars couldn't pass. There were turnouts about a mile apart where drivers halted, cut their motors, and listened for an oncoming car before proceeding. Once Mrs. Rose and her husband were caught at a tight place and had to unload their Ford, then hold the rear end over the shoulder of the road to make room for another car.

"When things got dull we worked up a neighborhood feud," she recalled of the Depression years. "We were all young, broke, bored, and scared, so it didn't take much to start a fight. Spring came at last, so we left Valsetz. We were glad to get back to civilization. Since then I have realized that while I may have left Valsetz, Valsetz never left me."

It's still thought to be "an ideal place to raise children" who must develop their own amusements, as they have always done, from playing cowboys and Indians to walking the taboo logs in the huge mill pond. But one brilliant and talented youngster found another way to occupy her time, which brought the world to her isolated doorstep.

One day in 1937 Dorothy Ann Hobson was seated beside her friend and benefactor, Herbert A. Templeton of the Valsetz Lumber Company, in the huge dining hall where daily the nine-year-old girl took her meals with burly loggers and sawmill men. Her folks ran the cookhouse for the Cobbs & Mitchell operation.

"There's going to be a newspaper in Valsetz," she whispered to Templeton.

Templeton read through the rough penciled draft of Dorothy Ann's first issue and quickly recognized the work of a youngster of unusual ability, charm, and mature depth. He made a deal to be her publisher. He would have his office force type, mimeograph, and mail her paper without changing a letter or mark of punctuation, and with no editorial guidance from him.

In her Christmas issue, Dorothy Ann announced the policy of the *Valsetz Star,* that "we believe in Fir, Hemlock, Kindness and Republicans," the capital letters hers for emphasis.

"Remember, this is a Republican paper, even if nobody listens," she bluntly warned her readers at a time when Franklin Roosevelt was at the height of his popularity and the Republicans in disgrace over the "Hoover Depression." The fact that anyone would speak out for the Republicans, even a young girl, drew more than merely local attention.

In issue after issue, the *Valsetz Star* dutifully reported the comings and goings of the local people, together with editorial comments of what was happening on the local, national, and world scene. The style and outspoken honesty of Dorothy Ann's words in her "Editor's Corner" captured the fancy of people across the country, bringing her and the isolated community attention that she accepted unabashed, with candor. Eleanor Roosevelt quoted her in a press conference. Dorothy Ann had letters and telephone calls from such notables as Herbert

Hoover (who had spent his boyhood in another Oregon small town), Big Jim Farley, Wendell Willkie, Al Smith, Shirley Temple, and Kate Smith, to mention just a few. Dorothy Ann's was a blithe spirit that captured the imagination, especially for her sympathies to Republicans, who found little welcome comment those days in the public press.

From a mailing list of about 100 prominent lumbermen, friends, and lumber company customers, the subscription list grew to embrace most of the states, the territory of Hawaii, the Philippines, and foreign countries, including Germany, with whom the United States wasn't on the best of terms. People were looking for something to take their minds off their troubles. Dorothy Ann's mail became so voluminous that the tiny Valsetz post office could hardly handle it; and the remote, rough-shod logging community didn't know what to do with a young celebrity in its midst who had put the town on the map and went merrily on her way, unaffected even after she was made marshal of the Grand Floral Parade of Portland's annual Rose Festival, attended a Republican gathering at the home of U. S. Senator Charles L. McNary, and spoke before the prestigious Portland City Club. Her writing and popularity seemed to parallel that of another Oregon youngster a generation earlier in another Oregon lumber town—Opal Whiteley, who as a child published a sensational national best seller, *The Story of Opal: The Journal of an Understanding Heart,* making you wonder just what there was about isolated logging and lumber communities that brought forth fresh inspiration and talent while writers and artists wandered across the world seeking the right conditions. There was some mystery about Opal Whiteley and her tale. But there was certainly no question about Dorothy Ann Hobson, whose pen was hewed as genuinely fine and honest as a logger's ax.

"We had a letter from Glen Rice of Kansas City," she wrote. "He sent us a dollar, but forgot to enclose it. He is a democrat." (She purposefully low-cased the party name.)

Having read the item, Rice came through with the subscription price, and Dorothy Ann commented:

"Mr. Glen Rice is a much nicer man than we thought. He sent us two dollars. We do not mind so much he is a Democrat. (This time she used the capital.) Besides, he may change. . . ."

On Senator McNary's reception, she observed: "You never saw so many Republicans eating salmon. . . ."

Dorothy Ann's spunky comments gave an unhappy nation many a needed chuckle. Mrs. Max Keiser, she announced, had been elected president of the Valsetz PTA. The *Star* observed:

"Mother thought she was going to get to be president, but she is only vice president."

Once Dorothy Ann announced about her father: "Henry Hobson went to Breitenbush Hot Springs to reduce and gained five pounds." And she noted on Henry Thiele, a prominent Oregon chef who handled the food for the McNary affair: "His stomach is the same size as daddy's."

After Shirley Temple wrote to her, Valsetz was never quite the same. Many subscribers asked Dorothy Ann to describe what life was like in the rainy lumber town.

"There are 600 cars, 300 refrigerators, 55 davenports, and 1 Sunday school," the *Star* reported. "One fortune teller; one beauty parlor, no picture shows; no bank—but mother and Mr. Grout will cash your checks if they aren't too big; one restaurant—daddy; two unions; one glee club; one post office with venetian blinds; 98 circular heaters; four strawberry patches; one bridge club; one barber shop; one newspaper (the *Star*); one sawmill; one handle factory; two timekeepers and a half; four dressmakers; 18 cows and no beer parlor. One doctor; 800 radios; one library; one baseball club; 600 Republicans and 22 Democrats; one CCC camp but it isn't here. yet. One high school, one grade school; six teachers with no pay and 500 dogs."

Yet despite the confinement of Valsetz, the *Star* had an amazingly broad view of national and world affairs, which caught the imaginations of readers across the country. Things that were left unsaid by the big dailies showed up in the *Star*. About France, the editor commented:

"They have many beautiful cathedrals, but they owe us some money. . . ."

"We stand for Peace and Republicans," she stressed in another monthly issue.

"We had a Republican meeting and nobody came, but we think we will have a bigger crowd next time."

Visiting Senator McNary's farm on a motor trip to the "outside," Dorothy Ann reported she happened to see "three democrats" stealing the senator's filberts.

"They ran when they saw us," she added with contempt.

Feeling that she may have moved too far into one political corner, the editor announced: "We will not say so much about Republicans in this issue because we had some papers left over last month."

The condition of the world and the country nevertheless greatly worried the young editor, who was showing potential for a lifelong career in journalism. On the war in Europe, she wrote:

"When we started the *Star* 2½ years ago, we believed in Republicans and Kindness. We still believe in Republicans but we have stopped being so kind on account of collections and Hitler. It's June and there is still no peace in the world and it's hard to be funny and cheerful, but we don't think Hitler will win everything and we think the Republicans will win in November, and we will keep publishing the *Star* because subscribers want us to. . . ."

She planned a blistering article on Hitler, written by her nine-year-old assistant editor, but he never got it done, even though he pledged if Hitler came "over to get us," he would throw away his glasses and fight.

"If we had a war," the *Star* reassured its readers, "Hitler couldn't get over the hill to get us because he would get stuck in the mud."

Still, at Thanksgiving the *Star* showed its kinder side when the editor announced that they would lay off Hitler for now, because "we can't say anything bad about anybody" at the holiday season. Then two months later, Dorothy Ann observed, "We won't say anything about Hitler this month because we are tired of him."

Dorothy Ann found the *Star*'s subscription list continuing to grow along with her personal fame. She was quoted by editors throughout the country, was interviewed on network radio shows, and met celebrities in Hollywood and on the East Coast. Politicos wooed her favor. None of this changed the viewpoint of the *Star*. But by the end of 1941 as the Japanese hit Pearl

Harbor, Dorothy Ann, who was now 13, had grown weary of the monthly grind and that Christmas terminated the *Star* forever:

"We've had lots of fun in the last four years, and we hope we haven't hurt anybody's feelings. We are sad about saying goodby. Mother is sitting in the living room, rocking and crying. . . . We are all crying now!"

The Hobsons left Valsetz, as most people do sooner or later. And the *Star* and its small blond editor were forgotten until a feature writer for the *Oregon Journal,* Jack Pement, discovered Dorothy Ann married and living near Stayton, Oregon. She had never intended to follow a writing career, she told Pement. She said so years ago before the Portland City Club, much to the consternation of newspaper editors who brought her there, believing they had uncovered the most unusual writing talent of that generation. And neither did she display any mementos in her home, nor had she kept any copies of her newspaper. Fortunately someone had, turning a collection over to the Oregon Historical Society following publication of Pement's articles about her long-ago activities.

North of Grays Harbor, the big and historic Simpson Timber Company operates Camp Grisdale in the Wynoochee Valley. Grisdale is the picturesque hub for some 200 loggers harvesting thousands of acres of fine timber in the Wynoochee, Satsop, and Canyon River watersheds, where wild man John Turnow roamed long ago, terrorizing the countryside.

Camp Grisdale, like Valsetz, also claims to be the "wettest spot in the United States," with some 12 feet of rainfall annually, which grows trees like jungle weeds, three feet a year. It is no wonder that Sol Simpson got into the tree-growing business successfully so early when the matter was still one of heated debate in the Big Woods. These lands were logged 50 years ago and are now being harvested again under a pioneering program begun in 1946, just a few years after the tree farms were introduced.

Joining with the U. S. Forest Service, Simpson's operates the nation's first *co-operative sustained yield unit* under a joint contract with the USFS for a 100-year period. The project contains 234,000 acres of Simpson land and 110,000 acres of Forest Service or public land, placed under intensive management for maximum production. The harvest in one recent year was 83,000,000 board feet, enough to construct 10,500 homes. Under the arrangement, there is a "working circle" of 270,000 acres, paced so the annual growth will exceed the annual cut feeding the Simpson sawmills, and plywood, door, and insulating plants in Shelton and McLeary. Some 80 railroad cars are shipped to Shelton each working day.

Grisdale, named for two of Sol's nephews, is a modern camp, although single loggers still sleep in their longjohns, even though they modestly don slippers and bathrobes when wandering about the place. The bunkhouses have rooms for single men, four to a room, and a bedmaker, who daily sweeps out the place. There's also steam heat, electricity, hot and cold running water, and modern plumbing. The men take breakfast and supper at the cookhouse, as loggers have done for generations, the meals costing about $.90. Some of these "single men" are married, in camp during the week and heading out on the weekends in buses with foam cushions. Thus logging camp life hasn't changed much in routine, save for the comforts that weren't even considered in the old days.

The object behind Grisdale is to keep the crews near their jobs and save the

time, trouble, and expense of a lengthy commute. It also assures Simpson of a
working crew on the scene, so efforts are bent to make Grisdale an appealing
camp.

When in full sway, Grisdale's population runs to 170 adults and 100 youngsters. The town has 43 family homes maintained by the company as a benevolent
landlord at a rental of $27.50 per month, including lights, water, and main-
tenance, far cheaper than a family could rent in almost any town in this day.
And the home lots are individual, with garages, lawns, and garden space. The
houses are heated with oil, which is also used for cooking, since the surrounding
trees in this beautiful setting (elevation, 700 feet against the southern Olympic
Mountains) are far too valuable to fell and buck for stove wood. But the over-all
aim is to make life in Grisdale "normal" for any small town. Housewives shop
at the Lumbermen's Mercantile, modern successor to the company store, but at
prices the same as in Shelton. Youngsters attend a local grade school, while teenagers ride the buses to accredited high schools in the vicinity. To stave off any
boredom, the town has its own recreational activities, including a bowling alley,
and there are community church services. The village gives off an air of permanency, and under the hundred-year cycle of this workable sustained yield program, Grisdale will likely stay at this place for many years. But it's still a company
town, governed by Simpson, and devoted to a single industry; and despite the
outward appearance of permanency, mapmakers will one day have to revise their
charts, for Grisdale's buildings are on skids so that the entire town may be moved
to another site, nearer to where the future harvesting will be taking place. And
even Sol Simpson's snub-nosed logging locomotive, which stands in the public
square, will likely be going along.

East of Grisdale, far out on the arm of fingered land bordering the western
side of Puget Sound, Port Gamble remains alive and well as another style of company town, unique both to the Pacific Northwest and the nation. The Pope &
Talbot people, that unusual family company which extends far back to colonial
times in lumbering and shipping in Yankee New England, are striving to preserve Port Gamble as a sample of what life was like in their conservatively operated company towns during the nineteenth century, when they began cutting the
original timber around Puget Sound.

Port Gamble was one of the starting points in the Big Woods, where the green
bonanza was felled and sent to market. At this very place, near the mouth of
Hood Canal, W. C. Talbot, A. J. Pope, and Cyrus Walker came ashore in 1853
to establish one of the sound's first logging and sawmill operations. There's a
bronze marker designating the site. They brought Maine lumber around the Horn
to erect the first buildings. The sawmill, operating by autumn, has been cutting
boards ever since, save for periodic shutdowns for improvements, expansion, and
repairs—the oldest continuously operating sawmill in the United States.

Thus the old town, patterned on East Machias, Maine, from which the company men and workers came, is still occupied and remarkably well preserved. It
isn't a ghost town by any means, for employees live in the many private homes
along the steep hillside above the sawmill, and the general store (established in
1853), the post office, church, and other public buildings also remain in use.
There is a distinct New England charm about Port Gamble, with its great elms

coming all the way from Maine to line the quiet streets, its Yankee architecture, and the beauty of the setting above the blue waters of Puget Sound.

Port Gamble gained in 1966 the distinction of being designated a National Historic Site for its "exceptional value in commemorating or illustrating the history of the United States." Since then, Pope & Talbot has been spending a bundle restoring and refurbishing the buildings, bringing back their original colors and style from under many layers of paint and years of neglect. The effort is the personal project of Mrs. Guy Pope, wife of the P&T president, who works with the resident manager, Charles Peck. The buildings are being specially marked with neat hand-lettered legends, telling something about each. Visitors discover that President Rutherford B. Hayes once called here to be dutifully honored with a gala reception; that the paymaster issued wages in silver dollars shipped from San Francisco and hauled unguarded up the street from dockside by wheelbarrow; and that the first U. S. Navy casualty in the Pacific Theater gained the distinction during an Indian skirmish when he "poked his head above a log to view a brave just fired upon and wrote himself into history." A museum of old photographs and artifacts is in the making on the second floor of the store, and another idea is to reopen the old theater of the community building, perhaps even showing early silent films.

"The cameras are still there," revealed Charles Peck, pointing to the high place where the old equipment is located.

Port Gamble has the potential of preserving one distinctive style of Northwest lumbering life on the wilderness frontier that was far different from the brawling, rustic logging camps and lumber ports normally associated with the old-time logging scene. Port Gamble had no saloons, no skid road, for the company wouldn't allow them. The only saloon was outside the town's boundaries, beyond company law. Workers journeyed across the sound to blow 'er in, for Port Gamble life was considerably more sedate and dignified, and activities on the conservative side, in keeping with the Yankee heritage. But it is an important part of the story, as is neighboring Port Ludlow, 10 miles away across the mouth of Hood Canal, over roadway through new timber on land that never went for taxes.

Pope & Talbot has held onto the historic site of Port Ludlow for another kind of modern timber town, a condominium development on the land of the old lumber village, which became Gamble's sister city. Nothing remains of old Ludlow save for the foundation of the sawmill's great burner along the waterfront, a few photographs in the restaurant, and the hills, which look the same as in early pictures. They call the private club The Admiralty, located on the very spot where Cyrus Walker had his plush private mansion. The grounds are beautifully landscaped, there are tennis courts, boat docks, walking and biking pathways, and comfortable private facilities for a complete escape from the clatter and nerve-shattering racket of modern living. Fortunately, Ludlow is off the major air routes. Seldom does plane noise cut the peaceful setting. The Pope & Talbot private condominium, where special guests are entertained in the style of modern business enterprise, is comfortable, functional, yet not overwhelming in plush or put-on decor. Its greatest asset is an overwhelming view of the sound, looking north, where big ships ply the Straits of Juan de Fuca and gray naval vessels swing south into Hood Canal toward the Bremerton Navy Yard. From the veranda, it is easy to imagine what the sound was like when those early lumbermen poked along its bays and inlets, seeking the right place to set up a mill and to drop those giant trees directly into deep water.

Port Ludlow is as geared to the 1970s as Port Gamble is historical of the 1870s, a few miles apart but in sharp contrast to each other. Both belong to the new age in the Big Woods where, more and more, the Northwest's timber people are growing to appreciate their heritage and at the same time are endeavoring to enhance the American style of the seventies and the future.

Chapter XV

Nobody Uses an Ax Anymore

The old-time logger was a fantastic breed of man. He did impossible things bare-handed, without any machinery at all. But the old logger died and passed on. The present-day logger is a businessman; probably has a master's degree in business administration and a B.S. in engineering. A B.S. in forestry. That's today's logger. The present-day logger would have lost his shirt back there. In the present-day logging crew you don't have anybody who does hand labor anymore. Everybody's got a machine in his hands. Nobody even knows what to do with an ax. . . .

ROBERT C. LINDSAY
Industrial forester
1972

The logger's ingenuity and the never-ending desire to make things easier, if for nothing more than to save his aching back, resulted in innumerable Rube Goldberg-style contraptions that evolved into solid tools of big industry.

The Big Woods seemed especially inspirational for maverick inventors—the brush was full of them—and the fact that the freewheeling lumberjack was a practical fellow, unbound by formal engineering education (until recent years) and technological do's and don'ts, placed him in a creative class by himself. What was readily available and worked best for him, from his view of the world about him, was perfectly acceptable until something better came along. And he was generally more than eager to share his idea with others. Get-rich-quick wasn't the prime objective, and in many instances ideas were copied and recopied without patent protection. My paternal grandfather, Joel Lucia, who spent much of his life in the Michigan woods, built numerous sawmills and was foreman for the Bay deNoquet Timber Company in Michigan. He was forever tinkering in his shop and came up with numerous "inventions," among them the sawmill log roller for handling the big stuff bound for the head rig, and a fire hydrant installed throughout the company town for quick action against sudden blazes so common to the industry. Both were widely copied for general use in the industry, but if the bearded old fellow ever got wealthy from his ideas, he must have buried the loot somewhere under a tree, for his family never knew about it.

Lumberjacks and lumbermen were constantly on the lookout for mechanical or technical gadgets applicable to the trade. That was how Joe Cox made out with the wood worms, where otherwise he might have passed them by, as thousands of others had done. It has been this way from the time the first lumberjack felled the first tree and then tried to figure out how it might be done easier. The Paul Bunyan legends were built on that same foundation; and it all reached an epitome in the Big Woods and in the California redwoods, where the jobs were bigger and tougher than encountered anywhere else. It's still going on today in the full mechanization of the forests, for once the logger and lumberman found ways to apply energy, from steam to diesel, to the damndest tasks ever confronted by ordinary men, and thus save their muscles and increase their comforts, there was no holding them back.

Ed Stamm, a logger for Crown Zellerbach, made a trip to Chicago about 1934 for some unknown purpose. Perhaps it was merely a holiday. But his logger's viewpoint never left him in the Windy City. Happening upon a municipal dump, Stamm watched in fascination while a tractor rig spread out the refuse with a huge sheet-metal plate bolted to its front end. Gasoline-powered tractors had been in the woods for years; cat skinning was an accepted part of the logging trade, but this was something new. Stamm mentally applied the dump machine to the Northwest woods. A stronger version of this blade attachment, including a vertical movement, might work far better for constructing logging railroad grade than the dragline scrapers and steam shovels being used in logging operations. Perhaps it would also work to handle slashing on the logging shows.

Returning to Portland, Ed Stamm got in touch with his friend Ernie Swigert, founder of the then-new Hyster Company, which would grow into a leading producer of logging, lumbering, and construction equipment in the decades ahead. Would Swigert build such a rig for Stamm? He certainly would, and he soon produced a pilot model, the first bulldozer, for use in the woods. Others wanted them, too, and from this logging idea evolved the tough, versatile earthmover that has become a basic tool not only in the forests but in heavy construction, freeway building, land development, on the farms, and in the front lines for Uncle Sam's wars since the island hopping of World War II, when there was a need for quick construction of airstrips to defeat the Japanese.

In the woods, no lumberjack worth his salt is ever content with the status quo. No sooner was steam king of the tall timber than the woodsmen began casting longing eyes at the internal-combustion engine and the possibilities of using electrical power. Part of the program of the Pacific Logging Congress one year was built around the question, "Will Electricity *Really* Work?" There were some successful electric yarders, but in about 1923–24, lumberjacks attending the Congress were considering the diesel as "having potential." Within a year, the Washington Iron Works was manufacturing a yarder, and the first diesel logging engine and shovel were being built, which would eventually lead to another revolution in the woods. Moreover, by 1970 it led to the full development, with hydraulic lifts, of some of the most powerful and unbelievable log-handling rigs ever conceived— machines that would grab a dozen or more logs like matchsticks, and yard, stack, sort, load, and otherwise tame them with such finesse that old-time loggers, were they to come back, would believe they had been reborn into some kind of mythical wonderland. What this huge and expensive gear accomplishes today is well beyond their wildest dreams as expressed in the Paul Bunyan tales.

You don't see the red hats, knit caps, plaid mackinaws, and frayed blue jeans that you did once. Loggers must wear hard hats now, under modern safety regulations, and they look more like construction workers than woodsmen. And when they go to town or gather at their conventions, they are largely in business suits, collared shirts, and four-in-hand ties, and only occasionally, perhaps as a salute to sentimentality, will someone break out in full regalia of yesteryear. But beneath it all remains the friendly, approachable, and freewheeling character of the logger; and when he bellies up to the bar, you know for a fact that he can still set a roaring pace extending far into the night. And he retains a fascination for the unusual, like one I recall who rambled on at great length about a certain watering hole "that's been modernized" and now contains some 2,000 silver dollars imbedded in the bar.

Simple, direct, practical thinking is retained as logging engineers, scientists, and foresters, all with college degrees, apply every possible Space Age technique and contrivance to the care, feeding, growing, and harvesting of timber. The wild, unbridled ways are generally gone, but the industry is still trying to live down Paul Bunyan's image with the public, which has caused it plenty of trouble—part of it certainly justified—in the pressure of the great ecology and conservation movement of the 1970s. The movement has forced changes in the woods that wouldn't have been considered either possible or economically feasible a decade ago, such as helicopter logging. But the public outcry against what is considered a raping of the nation's forests, coupled with a surging adult population that demanded more and better housing, have provided impetus to drastic change unlike anything since John Dolbeer's first steam donkey. The accomplishment itself is little short of amazing, since if ever there would be a staunch and stubborn holdout against the taming of the frontier's last stand, it would likely be found in the woods.

Logging and lumbering in the 1970s, and forestry, too, have moved with Jet Age swiftness away from the roaring times, so that while older members of the clan still harken to evidences of the past, with a fire in their eyes, the younger descendants of this brawny breed dwell on the present and the future. I recall coming upon a State Fair exhibit of old-time logger tools, everything from crosscut saw and ax to snoose can, whisky bottle, and peavey. An aging lumberjack and his wife were standing before the showcase and I hung back, watching, as he pointed out every tool, naming them all with joy and satisfaction, for they fanned an internal fire that would never entirely go out for him. The traditional gyppo logging operator is a fading hero of other times, replaced by tough yet well-trained technicians who know the answers without trial, error, and kill or cure. It has to be that way, for the rules and regulations applied now to the woods are such that high skill, training, and scientific knowledge are prime requisites to get along. Even timber cruisers use electronic calculators and computers to keep track of managed forests with the same attention to detail that Uncle Sam gives his taxpayers. The great, powerful rigs that get out the logs represent an investment of $100,000 and up, so that the independent gyppo logger and his banker look twice today when investing in any logging show. It's nothing like yesteryear, when a lumberjack just went out and located a good piece of timberland or secured a contract, put in his crude roads, felled the trees, dragged out the logs, burned the slash, and then walked away from it.

Logging the Big Woods today has become an exacting science under the constant scrutiny and patrol of environmentalists who keep the pressure on big and small timber companies, and even the owners of small farm tracts, to see that the loggers obey the laws, that impact studies are made, that good forest practices are being applied, and that there will be a follow-through in reforestation. The strong feeling is that the forests are everyone's business, whether public or privately owned, and that the public has a right to be heard and to observe.

Preparations for logging a certain timber tract require weeks, even months, of preliminary planning, mapping, aerial photography, consultation with foresters and game commission people, and even public education as to the nature of the project. Logging roads must be built with planning and caution, costing some $75,000 a mile, for roads into the back country are now considered one of the leading causes of soil erosion. And there must be a thorough cleanup of the logged-over sections. What was once burned as slash is now of value as wood fiber for pulp and paper, and then the land must be made ready for replanting again.

Modern power saws fall trees close to the ground, not abandoning those tall stumps with their springboard notches, on which old-time jacks balanced themselves. Every foot of timber is a bundle of dollars these days. The green gold is truly a bonanza, since single logs are worth over $100 apiece. Time is also money. Modern machines moving through the woods like giant prehistoric monsters can fell, buck, and load timber in what is almost a one-man operation and faster than an old boss logger could utter his favorite string of four-letter words. Comfort is the thing, too. Loggers sit in warm cabs, confronted by a collection of pushbuttons and levers to manipulate their rigs with skill and equal daring, listening for their orders to come via shortwave radio rather than for the gusty bellows that frightened birds into flight and sent wildlife scurrying for cover. Also, they are furnished with cassette tape recorders for background music on the job. It's no wonder that few want to abandon warm cabs for the wet and cold of the lowly choker setter down in the mud, for this still remains a tough, dangerous, and thankless job, which only those who enjoy the rugged challenge are willing to tackle. But now he, too, is being replaced by radio-controlled grapples that not only perform more efficiently but also may be operated at night, a distinct advantage in low-humidity weather.

Yesteryear's lumberjack dreamed of a skyhook to which he could attach his rigging and bring out the timber with the ease and grace of a flying trapeze. Just how strong was his longing for such a device was indicated in 1917, when balloon logging was suggested to the Pacific Logging Congress by R. H. Barr of Castle Rock, Washington. Barr's idea was to utilize surplus balloons or zeppelins of World War I, which were being employed to spy behind enemy lines, and equip them with winches instead of the stationary donkey at the landing used in modern balloon logging. Otherwise, the rigging layout and other details were the same, which means that Barr was far ahead of his time.

Balloon logging today is one of the modern methods for bringing out the sticks with as little damage to the site as possible. It has been practiced successfully for years by the Bohemia Lumber Company in the steep Cascades east of Cottage Grove, Oregon. Using regular round balloons and also triangular-shaped "sky-

57. Old Port Gamble on Puget Sound today is designated a National Historic Site. Still a company town of Pope & Talbot, the New England-style village has been restored and historic places marked for visitors. The plant in the foreground is the oldest continually operating sawmill in the United States. Note new forests growing behind the town. —*Pope & Talbot.* (Chapter XIV)

58. Modern loaders can easily loft a ton of logs in a single bite for stacking on a truck, or dumping in river or at mill cold deck, without the fuss of rigging. —*Western Timber Industry.* (Chapter XV)

59. In the good old days of yesteryear, systems of blocks, cables, and the traditional spar pole yarded the logs and loaded the trucks. —*Ellis Lucia photo.* (Chapter XV)

60. Great hydraulic rigs like this grapple loader handle timbers like matchsticks for loading trucks at the landings. —*Western Timber Industry*. (Chapter XV)

61. Rubber-tired skidders are the latest of the long line from bull teams to 'dozers or "cats." Ecology is the watchword, and the tires are believed less harmful to the land. —*Caterpillar Tractor Company.* (Chapter XV)

62. Even sawmills have changed. This portable minimill being demonstrated at the Pacific Logging Congress rough-cuts lumber at high speed and can be packed about by two men. —*Ellis Lucia photo.* (Chapter XV)

63. Sawmills and wood-manufacturing plants are far different from the rustic, noisy, smoke-belching mills of a few years ago. This is Pope & Talbot's new multimillon-dollar, computerized plant at Oakridge, Oregon. —*Western Ways photo, Pope & Talbot.* (Chapter XV)

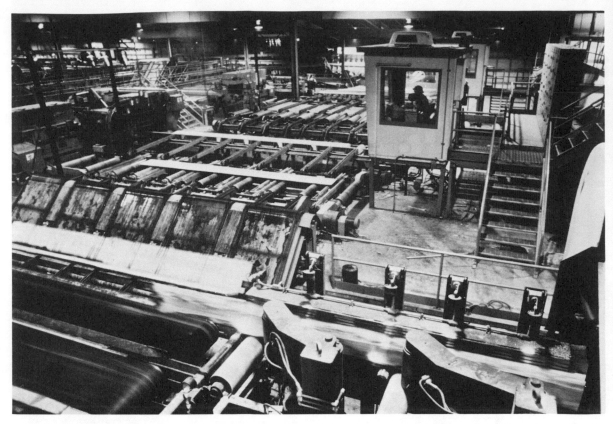

64. Modern wood-manufacturing plants are well-lighted, airy, and operate with computerized controls (at right), as exemplified at Pope & Talbot's new Oakridge plant. —*Pope & Talbot.* (Chapter XV)

65. High-speed machines do the work once handled by brawny loggers and mill hands, such as barking this fine log for plywood at Georgia-Pacific's Springfield plant. —*Georgia-Pacific.* (Chapter XV)

66. Veneer is stacked and clipped with efficiency in plywood plants like this well-lighted mill at Springfield, Oregon, a far cry from the handling by hand of yesteryear. —*Georgia-Pacific*. (Chapter XV)

67. Today's wood-products manufacture must consider its effect on the environment, which is a top-priority concern in the Pacific Northwest. This was a major aim when Crown Zellerbach invested many millions in Wauna, the old sawmill site on the lower Columbia River. Several mills in one, the $110,000,000 Wauna complex is highly automated in the manufacture of paper products. A fortune was spent reducing the odor that surrounds paper plants. —*Crown Zellerbach*. (Chapter XV)

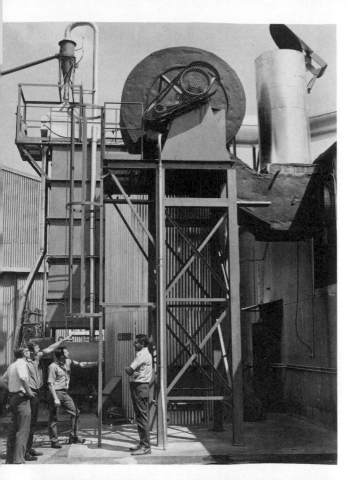

68. Like many Pacific Northwest wood-manufacturing plants, Boise Cascade at Albany spent a fortune on special equipment to reduce plywood pollution, winning an environmental citation. —*Boise Cascade*. (Chapter XV)

69. As wigwam burners were to sawmills, towering spar poles were the permanent landmark of the woods operations. But no more! Now mobile steel poles are moved from place to place as part of the logging equipment. Loggers call them tin spars. Space Age loggers yard the sticks in various ways, often with high leads and sky hooks, which are radio-controlled and can be stopped and started anywhere along the line. —*U. S. Forest Service*. (Chapter XV)

hooks," the system has proved a solution to the high costs of logging roads and the handling of timber in sensitive areas, or where it is difficult to harvest on steep slopes. Logs may be lofted for a haul from a few hundred feet to about a mile, cutting down on logging road extensions, soil erosion from skidding the timber, and damage to a new or young growth—all of which have become of much concern in the Northwest to the public, foresters, and lumbermen. It reduces log breakage, too, by flying the timbers out with the greatest of ease at higher speeds across rough terrain, swamps, steep slopes, rivers, and other natural obstacles.

The balloon system is rigged with a main line and haulback, operated from the log landing. In working a steep hillside, with the landing below, the haulback line pulls the balloon, filled with 530,000 cubic feet of helium, up the hillside on its own lift power and the guidance of the haulback, controlled by the stationary equipment at the landing. Loggers hook the tagline to the chokers when it reaches the top; then, as the haulback slackens, the balloon makes its lift and is returned by the main line to the landing, where the turn of logs is deposited.

The balloon with rigging weighs from 5,000 to 6,200 pounds and is approximately 113 feet high and 105 feet in diameter. The big Dacron bag can remain airborne for about a year with its capacity of helium. The sea-level lifting power of 25,000 pounds is enough to transport wood for a three-bedroom house in five trips, but being sensitive to elevations, declines about 5,000 pounds when working 5,000 feet above sea level. The rig has other uses. When being shifted from one site to another, crews are transported by gondola car lofted by the tagline. This is considered much safer than being taken by bus along a rutted, narrow, and precarious mountain road. So effective have been Bohemia's balloon logging operations that in 1973, the U.S. military was studying the feasibility of the method for loading and unloading ships.

Yet by 1973 helicopters were performing circuslike acts deep in the woods, lofting heavy turns of logs high above the treetops in what appeared to be the most versatile skyhook yet devised. Helilogging was beyond the experimental or "play" stage in both the Pacific Northwest and California. It wasn't exactly a new idea, but those who might have considered logging with whirlybirds a decade ago had quickly cast the notion aside, since the choppers simply were too unpredictable, not powerful enough, and expensive to operate. But times changed radically, and a combination of demand for more timber (there was a critical lumber shortage in the United States in 1973–74) plus environmental pressures brought helilogging into full sway through its use on federal logging shows, with encouragement by U.S. foresters.

Much of the pioneering was by Wes Lematta, the barrel-chested founder of Columbia Helicopters, Portland, who has made a career of doing unusual things with whirlybirds. And by 1973 his outfit was still a leader in the work, one of four in the country, if not the world. The public first heard of Lematta in 1957 when with skill and daring he plucked 15 men from the sinking dredge *Rosell,* which had been rammed by a Norwegian freighter near the entrance of Coos Bay, Oregon. Lematta picked them off, one by one, amid the guy wires and high wind, and in the poor light of dusk. The deed—"something that I had to do"—brought citations from the U. S. Coast Guard and the Department of the Army, not to mention the words of thanks from a grateful crew, which otherwise would have drowned in the onrushing tide. Wes went ahead distinguishing himself with his varied skills as a 'copter pilot in sea rescue work with Fred Devine, the West

Coast's legendary King of Salvage, who utilized logging techniques on his famed *Salvage Chief*. Lematta turned his machine into an "aerial crane" to erect steel power towers, string power lines, set poles and pipelines, and carry tons of equipment to inaccessible places, demonstrating that the chopper could be used for many purposes besides rescue.

Helicopters were becoming larger and more powerful all the while. It seemed natural that Lematta and his bold crew would become interested in the possibilities of logging by whirlybird. Logging and forestry eye everything as a possible tool, and the helicopter was certainly no exception. It had been hovering over the woods since the first days of the Tillamook Burn reforestation project, when a small 'copter was used for aerial seeding and also the construction of a mountaintop lookout. Always the question was, "How can we put this to work?" Economics has a lot to do with it, and the high cost of chopper operation made it prohibitive. But the winds were shifting rapidly. In 1967, a decade following the *Rosell* incident, Columbia Helicopters, which by then was heavily involved in various construction projects throughout the West Coast and even in the South Seas, acquired a Sikorsky S-61 with an external load capacity of 8,000 pounds. The rising value of logs plus growing public concern for the environment, especially in public-owned forests, made yarding by 'copter appear now more than a stuntman's dream. Logging costs were going up all the time, and one of the prime expense items was road-building.

A test run was made in January 1971 on Drum Creek in California's Plumas National Forest, where a timber sale had been awarded the Erickson Lumber Company of Marysville. Terrain and distance problems were such that logging by chopper appeared the answer to the difficulties. Lematta and company lofted 4,000,000 feet of ponderosa and sugar pine, and while the financial outcome was "marginal," use of the whirlybird was considered a successful pioneering adventure. From that day, there was a steady surge of interest in helilogging, encouraged furthermore by development of more powerful machines for use in Vietnam and which were later being used in the Big Woods as more and more surplus ships became available. Columbia expanded its fleet to eight machines, adding seven Boeing Vertol 107-IIs with sling load capacities of 11,500 pounds. Erickson also got into the act with a D-64-E having a maximum capacity of 20,000 pounds, while an eastern outfit, Carson Helicopters of Perkasie, Pennsylvania, became a third pioneer, using twin S-61-L aircraft with 8,000-pound capacities apiece.

The helilogging system is basically what logging has always been, except that there are special considerations, such as the weight of the logs, a judgment that chopper crews must trust to the men on the ground. The typical specially trained crew of a 'copter logging show includes fallers, markers, chokermen, hookers, a landing chaser, and a loader operator, in addition to pilot and copilot in the air. Fallers must be carefully trained in judging local timber to consider the chopper's weight limitations for lofting the stuff to the landing. The markers are also all-important in judgment. This isn't as simple as it sounds, since log true volume and board foot content are imperfect, and therefore the log must be judged by cubic volume, including the bark. Markers use volume weight tables based on 35 to 67 pounds per cubic inch. Here, economics enters the picture. Because of the high operating cost, each load must be as near capacity as possible, and split-second timing is necessary, since time is many dollars. It's largely up to the marker to make up the log turns as near the aircraft's maximum lift capacity as possible, using colored paint or tags to indicate to the chokermen which logs are combined for each turn.

Those chokermen and hookers are right down in the mud, as in the good old days, in the toughest job of all, scrambling over and under logs to beat many of the same brutal obstacles that have always plagued them. However, in helilogging, the choker setter works two to four hours ahead of the actual yarding operation, thus requiring from 100 to 300 chokers. Unlike highball days, the chokerman isn't paced by the tight three-minute cycle of the chopper. But the hooker still is, although now he keeps in constant radio contact with the pilot and copilot, as do others in the operation. Radio is a big thing in the woods now, having replaced the faller's and bullwhacker's lusty shout. In some ways, helilogging is safer than strictly a ground-level show, for the pilot has constant surveillance of the men, the loads, and the hangups, and can direct the crews to safe positions or warn them of dangers from his specially installed bubble window. From this viewpoint he looks directly down upon the lines and can maneuver the plane accordingly. His copilot meanwhile keeps a steady eye on the instruments, monitoring fuel flow, temperatures, and performance. It's grueling work, which requires heavy concentration, so the shifts are shortened to four to five hours to avoid dangerous errors in judgment resulting from pilot fatigue. The work is risky, as with all logging, but accidents are rare. Lematta went a long while without a serious mishap, but in 1972 lost a 'copter pilot and copilot from an undetermined cause. Overloading isn't a factor, however, since special safety devices signal an abort if a load is too heavy.

The hooker, decked out in bright flagmen's vest and red hard hat, is easily spotted from the air—a tough, woods-wise athlete able to scramble rapidly about the operation and make quick decisions as to the best way to lift a log turn without tangling the package. He must also contend with strong helicopter downwash. The pace is terrific, and on a rugged slope it takes a logger with the agility of a mountain goat to outrace the helicopter to the next turn before the plane completes its delivery cycle. More often than not, the terrain is very rugged; if it weren't, this probably wouldn't be a 'copter show.

Sometimes it requires two hookers to keep the pace, for one simply can't cover the ground, and you can't just leave a helicopter hanging up in the sky, waiting for the man below. It's too costly. The load line, attached to the plane's bottom but not to a winch, ranges from 50 to 225 feet in length, depending on the terrain. The plane itself serves as a winch. The pilot controls an electric switch-actuated cargo hook to connect the line and the chokers. The electric line runs down to the hook alongside the main line. In emergencies, there are three electrical and mechanical methods to jettison the logs and rigging should a flight crisis arise. On some very hot days, a crew may abort a load 15 to 20 times. For safety reasons, a flight pattern is worked out that is free of crews and equipment, in case the plane needs to dump a load.

At the far-end landing, the 'copter delivers the logs with chokers attached. The landing chaser gathers up the chokers, coils them individually, and places them in piles of 10 to 20 for the plane to return to the chokermen. The landing is usually within a mile of the woods operation, where there is a good road access. Here, too, there are special requirements, not just any knoll or clearing. Distance of the haul must be within feasible economic limits, and the location must be one that won't require too steep a climb by the heavily loaded 'copter. Wind conditions and flight paths have to be taken into consideration. Lumberjacks must also pay attention to their housekeeping, watering the landings in dry weather to keep the dust down and picking up litter, including paper, plastic containers, shirts, and jackets, so it won't be ingested by the engines on the rotor downwash. There must be space, usually rocked, for parking and refueling, and adequate area for fuel storage for a week's operation. The 'copters consume from 1,500 to 4,000 gallons of turbine fuel in a single eight-hour shift, and burning 500 gallons an hour, they can run low in 40 minutes. There must be a fully stocked maintenance van of parts and supplies, manned by two shifts of mechanics. That's why costs run so high, from $800 to $2,500 an hour. Lematta's cost runs $70 to $80 per 1,000 for yarding. That's reason enough why the contracts are figured in terms of minutes and seconds for getting out these logs.

Heated criticism by environmentalists and the constant debate over if, when, and how the nation's forests should be harvested have boosted the stock of helicopter logging shows. It's sort of an end-around-the-line answer to critics who object violently to clear-cutting in the back country of U.S. forests, particularly in fragile areas, and in the construction of more logging roads or skid trails, which are prime targets because of serious erosion problems. On the other hand, foresters and lumbermen point to quantities of good timber that will be wasted if it can't be harvested when the demand at home and abroad is at an all-time high. The U. S. Forest Service has thus been conducting special projects with the whirlybirds in the Northwest woods, bringing them beyond the experimental stage, as an answer to one of the critical problems of modern forestry. Since 'copters lift logs vertically into the air, damage to new and immature growth and the soil is minimal, therefore greatly reducing ground erosion. Small patches of diseased or overmature timber may be logged without building a high-cost road system. There are areas of sound, usable, 300-year-old timber that are inaccessible in the high mountains, where much of the federal forests grow, or where unstable soil on slopes or adjacent to streams or public scenic attractions can be aerial logged without serious damage. And weather conditions aren't as great a factor as might be supposed, for the 'copters can work in rain and snow, and in wind to 25 knots. In the summer of 1974 Lematta was successfully operating a nighttime helilogging show which promised to be an additional boon to the industry and foresters in low humidity weather. All things considered, the timbermen and foresters are looking toward the helicopter as one more very important tool to be used where conditions require filling the nation's mounting consumption of wood products by operations that might otherwise be economically or ecologically unsound.

Since the first Tillamook Burn experiments of the late 1940s and 1950s, the helicopter has found increasing jobs in the Big Woods. Aircraft of varying sizes are used for aerial seeding, timber cruising, photography, fertilizing, spraying, road surveys, mapping, fire patrols, and transportation into the back country. Like all else, this latter is an important factor, not only in time saved but also because hiking takes hours to cover the distance, for which the crews are paid. A 'copter may reduce several hours' walk time to a few minutes. Those who work in the forests today still love the woods and the outdoors with a passion, but things are no longer done for free, or for the joy of doing. Wages have become as much a factor in the forests as along an industrial assembly line.

Yet helilogging, unknown a decade ago, is merely one of an amazing array of Space Age magical changes in the full mechanization of logging and manufacturing plants, right down to computerized forestry, logging, and sawmilling. The strange, grotesque rigs appearing as prehistoric monsters from another planet, thrashing through the Big Woods, would undoubtedly hand old-time buckers and swampers such a jolt that they'd swear off the snakebite forever. Lumberjack ingenuity, as we have said before, has always been the salvation of the timber industry, but until now it has entailed a lot of personal muscle and brawn. Today, the big rigs with yawning jaws and steel claws that reach out to grapple the fat sticks in bunches and lift them from one place to another are maneuvered by a single skilled driver in a warm cab, sheltered completely from wind and cold, the rain and the snow; and the yellow-and-red machines remind you of some thrill ride in an amusement park or perhaps those toy grapplers of the penny arcade where for a coin, you manipulate a claw shovel over a bar of candy or special prize, which is then swung to a chute to drop into your hand.

A modern logging show may have $750,000 in massive machines, and even the smaller independents, successors to the old-time gyppos, are able to invest huge sums in diesel-powered, hydraulic lift grapplers, skidders, and loaders. Skyline logging "flies" the big sticks above the ground, eliminating damage to new growth, with the droplines extending from radio-controlled cars traveling along a 1.5-inch cable, where they can be stopped and locked into position with 150,000 pounds of pressure grip anywhere along the skyline. Huge yarding grapples have made nighttime logging possible, under lights, for these rigs eliminate the need for choker setters and hookers who can't work in the shadows of night. The good old bulldozer, which was spawned in the Big Woods and has long been a basic tool of timber mining, often serves as a power unit, as did the steam donkey and locomotive of yesteryear, to which may be attached giant tong grapplers, which have grip cable features and are mounted on swivels and hinges, and with the use of hydraulic mechanism, can be maneuvered into most any position to wrestle heavy logs from the woods.

Equipped with winches, the modern 'dozer can serve as a tension skidder, that all-important tool of any logging show; and many of them are designed so that they can be easily converted with "attachments" to angle 'dozers, straight 'dozers, tilt 'dozers, U 'dozers with winches, and hydraulic rippers to serve whatever conditions confront the lumberjacks. Many of these snorting machines, too, are equipped with rubber tires instead of tracks, again bowing to modern forestry and the recognition that better care must be taken of the earth. Operating the rigs takes special training, for the logger of today uses the muscles of his brain as well as the brawn of his arms, legs, and back. He needs to know what lever to pull, what button to push—and when.

The machines are every bit as big in size as yesteryear's logging gear, but far more powerful and maneuverable. Some are so versatile in their construction, every moving part able to operate independently of the others, that they can squat, bend, double back, and stand on their haunches like the log-loading elephants of India to get the job done. They do everything but climb a tree. One giant 55,000-pound skidder boasts a 70,000-pound maximum pull line from a live-shaft winch and is powered by eight-cylinder engines. The Ranger 880 was particularly designed for the rugged timber country of the Pacific Northwest and the California redwoods because of the selective cutting methods being employed in areas once excluded because of high logging costs. The 28-foot rig can double back on itself like a circus performer, with a turning radius just under 24 feet, and a 90-degree turning angle. Blade and winch controls are operated from the comfort of the cab at "armchair height"—says the advertising material—the driver riding on a bucket seat with armrests. A second 10-foot blade may be mounted on the skidder's rear to serve as a ground anchor when winching payloads up to 50,000 pounds.

For generations, spar poles—those towering stripped giants standing alone on the highest ridgetops—have been landmarks of the industry, as have the wigwam burners. Long after the logging operator moved on, the spar pole remained silently on the high hill. Now loggers are turning to mobile yarding equipment with steel spar poles on wheels, which can be easily shifted from one location to another. It also means that one more fine tree can be felled for the wood-pile. The mechanical "tin spars," which run to 160 feet in height, are mounted on four-wheel drive carriers, making them easy to move from place to place for thinning and yarding operations. Hydraulic log loaders that can handle upward to 100,000 pounds are amazing, alive creatures. One particular model, called the Berger Logloader, weighs 113,000 pounds, has a removable counter-balance of 9,500 pounds, and is able to handle 26,000 pounds at a 25-foot extension and 15,000 pounds when reaching out some 40 feet. The heavy-duty grapple, all hydraulic, has a continuous 360-degree rotation, even with ca-pacity loads and a yawn from 8 to 84 inches. It is quite a rig, yet typical of many now being offered the timber trade of the Northwest, with luxurious chrome-plated piston rods and nine-inch pistons overlaid in bronze to prolong their lives; and for the cab of heavy-duty half-inch plating, there is air con-ditioning and special window guards. The crawler system also has hydraulic, expandable tracks operating independently, so that the monster can turn, if you will, on a dime.

These fascinating log loaders and handlers can sort, load, unload, stack, and restack, juggling tons of logs like so many jackstraws. The entire technique, from woods to mill, has changed completely, and the mill pond itself is on the way out. Many mills now cold-deck their logs, for although water is a great preservative (logs can lie for years in rivers or ponds and still be sound), the cold-deck system enables the easy sorting, grading, and handling of the sticks, whereby in the ponds, the logs could easily become mixed. Again, it's a matter of efficiency and economics. The powerful loader jaws can grab 50,000 to 100,000 pounds of logs at a time, loading a truck in moments at the woods landing and with far less danger to the lumberjacks. Loading time is cut in half, costs by a third, and therefore it is no wonder that the loaders, which travel from landing to landing with the trucks, pay for themselves within a short time.

The log trucks, as huge as Shay locomotives, exhibit more muscle as the years advance, with diesel units of nearly 500 horsepower and 1,500-square-inch cooling areas able to haul many tons along the muddy roller-coaster logging roads. Their cabs are thickly insulated against noise and weather, are air conditioned, and sometimes are equipped with tape recorders for background music in addition to shortwave and standard radio systems for the enjoyment of the driver, who despite a highly responsive piece of equipment, must still be skilled at handling the heavy rigs around whipping curves. Another gadget

57. Old Port Gamble on Puget Sound today is designated a National Historic Site. Still a company town of Pope & Talbot, the New England-style village has been restored and historic places marked for visitors. The plant in the foreground is the oldest continually operating sawmill in the United States. Note new forests growing behind the town. —*Pope & Talbot*. (Chapter XIV)

58. Modern loaders can easily loft a ton of logs in a single bite for stacking on a truck, or dumping in river or at mill cold deck, without the fuss of rigging. —*Western Timber Industry.* (Chapter XV)

59. In the good old days of yesteryear, systems of blocks, cables, and the traditional spar pole yarded the logs and loaded the trucks. —*Ellis Lucia photo.* (Chapter XV)

being added to the log haulers is an electronic weight calculator so that trucks can be loaded to capacity but not outside the law. Time was when truckers ignored the weight limits set down by the states, for it was cheaper to pay the fine for overloading than to underload, but now in a cost-conscious age, where pennies become dollars, the need is vital to stay within the limits.

Skidders . . . 'dozers . . . haulers . . . mobile yarders . . . loaders . . . trucks . . . scrapers . . . tractors . . . on rubber tires or steel tracks, in the woods, along the haul roads, and at the landings . . . skidding, scraping, loading, pushing, pulling, cutting, clearing, stacking, piling, grading, building . . . all tied together by intercom and shortwave radio systems, which were once only luxuries of state and federal foresters. Soaring inflationary costs in wages and necessary equipment, the high price of stumpage, and the outcries of environmentalists to do a better job have spurred the timber operators to achieve, as never before, far superior methods for filling the woodbox, even at an investment in machines with six-figure price tags and which, in the rugged, rough terrain of the Big Woods, take a terrific beating, so that they wear out within a few short years. The combined conditions thrust the timbermen in the early 1970s off the launch pad into the moon-shot electronics-computer age with the thunder of a takeoff for outer space. There has been a kind of frenzy about it and a conversion that few would have thought possible a very few years ago. Space Age mechanization has become so widespread, so rapid, that the final capitulation of this traditionally rough-and-tumble industry amazes the mind. The shedding of the old ways is producing a new world that Andy Gump and Big Mac and the Polson Brothers and those brawny crews stretched from the Humptulips to Eureka wouldn't recognize much, be it a logging camp or a manufacturing plant.

"Nobody even knows how to use an ax anymore," comments forester Bob Lindsay of Crown Zellerbach. "You can hardly find a man who knows what to do with an ax."

A bit extreme? Not really, for what Lindsay means is that most everything is mechanized now save for the choker setter, and he may also be on the way out, replaced by grapple logging. And it doesn't mean that the timberman has gone soft, for whether he be logger or forester, plenty of rough tasks remain in the raw outdoors, and the risks are still there. You can still pick up the paper and read about some poor guy who was crushed by a rolling log. And loggers are still loggers, freewheeling, down-to-earth, and fun-loving, sufficiently so that the miniskirted hostesses of an airline used for commuting by lumberjacks up British Columbia and Alaska way complained about the pinching and asked if the company would allow them to wear something less vulnerable.

Since the days of railroading in the woods, it's taken more than a sidehill of timber, a misery whip, an ax, and a span of horses or oxen to get into the logging business in a big way. Now it costs a small fortune. The assortment of rigs ranges from small portable sawmills, which two men can carry right to the logging site, to those monstrous machines that yard and load turns of logs in great easy strokes. They come in such varieties, shapes, and price tags that when the Pacific Logging Congress staged its equipment show in 1971 at Portland's Memorial Coliseum, the bright and shiny rigs jammed not only the hall's interior, but much of the parking lot, an estimated value of $35,000,000 in logging equipment on display.

The aim is "full utilization," with as little waste as possible. It's become the lumberman's religion at long last. Stronger metals have led to thinner saw blades both in woods and the mills, reducing the sawdust waste. Some thirty years ago, people of the Northwest burned sawdust in their homes, and it fed the many steam power plants of the region. Today the huge sawdust mountains are a rarity, and much of what is there is turned into wood by-products or utilized as fuel within the plant. Wood chips and fiber shavings, bark, and sawdust are major items of today's timber manufacturing; what once went up in smoke as residue now goes into paper, packaging, fiberboard, particleboard, and even bark mulch for landscaping and lawns. Boise Cascade saw such potential in this wood waste use that it increased the capacity of two of its plants from 110,000 tons to 400,000 tons annually for the manufacture of pressed wood from shavings and sawdust for home and furniture construction, and even found a way to utilize quantities of bark in particleboard manufacture. Most all the major manufacturers have turned to this kind of operation, so that wood waste isn't waste anymore. Bohemia, Inc., for one, announced in 1973 that it would build a $2,500,000 specialty plant to convert fir bark into vegetable wax and cork adhesive extenders at a rate of 40,000 tons a year.

Full utilization has brought another dramatic change in the logging operation. Slash burning, which once was a commonly accepted method of cleaning up a logging operation, is now practically nonexistent, save where it may be virtually impossible to haul out the residue. The wigwam burners, which were long the industry's familiar landmark, are being slowly phased out, for what was formerly waste material now provides over half the pulp for the manufacture of paper in the Northwest. The tall stumps are no more, for with power saws and mechanized cutters, trees are sheared close to the ground; and stumps, knots, and burls also become a part of modern wood production. Full utilization: bark becomes mulch, insulating material, and fertilizer; sawdust and shavings are compressed into fuel logs for fireplaces and Franklin stoves, which will become increasingly important if the nation suffers a sustaining fuel shortage of oil and electrical power. Trimmings, edgings, and unused slab are turned into chips. One firm uses bark and sawdust to manufacture glycerine, nickel, and sulphate. In all, about 2,000 products come from wood, some of them far removed from building materials, such as the wonder drug DMSO (dimethyl sulfoxide).

Paper manufacture has been an offshoot of the timber industry since pioneer times, but the wood chip industry has now also become an important stepchild. Fleets of chip ships haul much fiber abroad, principally to Japan, as scientists continue experimenting with leftovers by taking the wood apart, dissecting it, and testing its adhesive qualities through soaking, drying, boiling, freezing, and applying tremendous pressure to turn out thin sheets of hardboard, strong as steel, which can be bent, molded, and shaped into hundreds of useful and attractive products. Chipping machines have even invaded the woods to take care of leftovers on the spot, and to tackle slash and brush, which become serious problems in the replanting of any clear-cut Douglas fir area. These machines would have found a bonanza in the Tillamook Burn cleanup prior to reforestation. The chipper's heavy knife blades eat small residue and even large logs and railroad ties, reducing them to pulp, mulch, and landfill. Another spiked roller rig transforms forest waste into mulch, returning it to the soil as part of the logging mopup to prepare for the regrowing of trees. It is somewhat ironic in that, as logging operators strive for ways to leave the soil as undamaged as possible, foresters have found that a certain amount of stirring up of the soil helps stimulate the new growth, so that the problems of silviculture seemingly have come full circle.

The Ecologizer, which the manufacturer describes as a "revolutionary portable small sawmill" is designed to produce dimension lumber from small logs on the spot. It will cut logs from three to 10 inches in diameter. With an eye on the ecology, the Ecologizer appears to be a clue to the future, while lumbermen are still harvesting virgin or old-growth timber. The growth of the replanted forests of today probably will never reach the tremendous size of those old growth firs, spruce, and cedar, which were standing for several centuries. It can't be afforded, for the demand for wood products is heavier than ever, and loggers and sawmills can now handle the small stuff (in fact, many have been retooled for it), for, since chips and wood fiber have come in so strong, it seems to make little difference whether the chips are from large or small trees. The Ecologizer, according to the manufacturer, produces no edgings, slabs, or chips, just a light sawdust blown out through an oscillating pipe and spread over a large area, as mulch might be spread on a garden. This helps prevent erosion, conserves moisture for reforestation, and is fully biodegradable. The boards, smoother than regular "rough lumber," are produced at an input rate of 100 feet a minute, from a 210-horsepower diesel unit. This fascinating small sawmill is only eight feet wide, 14 feet long, and 7.5 feet high, and is so compact that it may be easily moved from one site to another. It is completely enclosed with tight lockup panels as protection against vandalism—another factor of modern American life that has to be taken into consideration in the woods. But the fascination with this compact sawmill is surely its small size, when I look back nearly 25 years on a "portable sawmill" being used by a gyppo outfit in the Tillamook Burn, and which was considered quite maneuverable, although it had to be skidded on and off a huge flatbed or logging truck, far more cumbersome than modern midget mills that a few men can pick up and cart away.

Small, efficient, portable mills are just one more sampling of what has been going on in the woods in the 1970s. Now the industry is moving into the world of computers in the grand old American style, all out for anything new. Computers are being used for most everything, from timber management and fire control to slicing up logs in the mills. The day may come when the old-fashioned timber cruiser will be largely a thing of the past. He cruises ofttimes by plane now, electronic calculator in hand, rather than tramping through the woods and brush. And even when he does hike through the timber, his small computer goes with him.

Banks of computer data, like those stored by Crown Zellerbach, keep tab on the forests from cradle to grave, so that company foresters always know where they stand with any tree in their widespread holdings. Portable scanners on woods operations analyze a log before it ever reaches a sawmill or pulp and paper plant. Safety scanners ferret out chunks of iron, spikes, bolts, and even an old ax head hidden deep in the log. The Wobbly saboteur of yesteryear wouldn't stand a chance today. Automatic log sorters and photoelectric scanners, linked to programmed computers, rate each log for maximum recovery of its contents by the way it is sliced into boards. Again, the aim is to get the most from each costly log and reduce the waste as close to zero as possible. The guesswork of the old-time head sawyer, whose refined skill was indeed unique, is being replaced by data processing and programmed computers that spew forth all the answers on how to cut each log faster than a lumberjack can spit. Since no two logs are exactly alike (although industrial foresters hope someday to come close to growing uniform trees), this computerized calculating system is a kind of magic to veteran sawmillers.

The scanner, with a series of light detector arrays fixed above, below, and at various angles to the log, inspects the timber. Then an electronic control module converts the information into data compatible to the computer for simple and rapid cube volume scaling, plus a continuous readout of the log's diameter and any projections such as large knots, snags, and splits, so that the operator can instantly intercept oversized, undersized, or imperfect logs that may cause trouble at the head rig.

Pope & Talbot spent some $8,000,000 and three full years computerizing its big plant at Oakridge, Oregon, into a refined operation that would handle small and big logs, plywood peelers, and chips in line with what is felt to be the new age in the Pacific Northwest timber scene. Its complicated computers are so sophisticated that they can figure a log cut in line with instantaneous changes in the market value. Photosensors determine the maximum and best use of each log, which is then routed quickly to sawmill or veneer plant in accordance with the determination made at the log-processing center. As a log passes through the scanners, a portion of the light is blocked off on the detector array, and this is directly proportional to the log's diameter. The amount or degree of blocked light is "read," and this is fed directly to the control module and then to the digital readout unit, which has been preset according to whether the system is handling reasonably uniform logs, or sticks that are oddly shaped or tapered. The electronic information is converted into electronic data for scaling, sorting, bucking, and just how to slice-up the log. Such scanners are used not only at the headsaw, which is becoming a chip-and-saw technique, but at resaws and edgers to take the guesswork out of manufacturing, save time, and reduce the losses from poor log utilization.

High-speed chippers are now installed at the head rigs of many mills, ahead of the saw, to reduce knots, burls, and other log defects to chips in an instant, rather than having to transfer the outer slab to another part of the mill for chipping. It's all one automatic, smooth-flowing operation, although the sawyer has an override in case, in his judgment, the computer has made a mistake. This is characteristic of the lumber-logging fraternity, where the final trust still lies with the man at the switch. But generally, based on size and taper, the computer system searches through over 1,800 cutting patterns for the best method and saw setting for each individual log.

High-speed operation and small saw kerf are other factors in efficiently processing logs of a minimum four-inch diameter at the P&T Oakridge mill. It is estimated by President Guy B. Pope that the new mill will develop about 55 per cent more lumber from each log. In addition, the company in 1972 signed

a 10-year, $35,000,000 chip contract with the Japanese so that little will
remain behind as waste when the mill is in full operation on 42,000 acres of
old growth in the area. Even so, it will likely take 20 years to cut through the
stand; then the plant will turn to federal timber until the new growth comes
back and is ready for another harvest, this time utilizing smaller trees. The
primary aim used to be growing *big trees,* but in recent years the value is
such that "we can't hold onto old-growth trees," says Guy Pope. "If you take
the interest on investment and deduct taxes, it isn't practical for us not to
cut the stand and replant." The company now considers its course in terms of
earnings per share, and therefore has in five years rebuilt all of its manufacturing
facilities to handle the new world of forestry and lumbering.

Space Age sawmills and wood manufacturing plants are well-lighted, modern, safe, and very fast, able to slice up 1,000 logs in a single shift. They are in
sharp contrast to the plants of yesteryear, some still around, which buried
themselves in the waste of inefficiency and guesswork, and were so unsafe and
ill-lighted that crews feared for life and limb, with heavy justification on every
shift. Ranking high among the industry's sleek new plants is the paper and
pulp mill complex of the Crown Zellerbach Corporation at Wauna, site of a
historic sawmill operation on the lower Columbia River in the early decades of
this century, when the big cut was shifting south from Grays Harbor to the
Columbia show.

Paper making has been an integral part of the Northwest timber scene since
pioneer times when, in the midst of wood a-plenty, settlers and business tycoons
had to rely on shipments from back East for their paper supplies. This now
seems almost as ridiculous as the story of one pioneer home, still standing at
Oregon City, which was prefabricated in Maine, and how Pope & Talbot
carried boards all that distance around Cape Horn to establish their first
sawmill and company town at Port Gamble. It was Henry Pittock, one of
Portland's early civic leaders, who pointed the way to the paper industry.
Pittock had gone to work for the weekly *Oregonian* and then later bought the
newspaper. At times a lack of paper almost killed the publication, especially
when Pittock stepped it up to a daily to publish telegraph reports on the War
Between the States. Then, too, paper arriving via Cape Horn or across the
plains was often badly damaged.

Pittock realized something must be done, for it was like the old adage of
"water, water everywhere and not a drop to drink," with the forests in
abundance all around him. But he knew little about paper manufacture when
he set up a mill in 1867 at Oregon City. Neither did his partner, one William
Wentworth Buck. The operation was a dismal failure, but this didn't dampen
Pittock's enthusiasm and his need. He secured a new partner in William
Lewthwaite from California, who was an experienced paper maker and had
performed successfully with a pioneer plant north of San Francisco. This time
the Oregon mill was established at Park Place along the Clackamas River
north of Oregon City. Its principal raw material would be straw from the
wheat farms of the Willamette Valley to the south. The mill could also utilize
manila rope, old fishnets, and discarded ships' sails, and by 1870 was recognized as the first permanently operating paper plant in the Pacific Northwest. Later, Pittock, who became a celebrated early-day newspaper publisher,

built a larger, flourishing plant at Camas, Washington, only a short distance from where the Hudson's Bay Company operated that first sawmill long ago.

The companies that evolved into Crown Zellerbach moved into the Pacific Northwest after the turn of the century with mills at West Linn, Camas, and Lebanon, and sales offices in Seattle and Portland. Pulp and paper manufacture became an important offshoot of the lumber manufacture and the original recycler of waste material discarded by the lumbermen. Under the current scheme of things, the paper companies, of which Crown Zellerbach is among the nation's biggest, have become major users and exponents of industrial forestry for the future, planting and planning for many decades ahead, and not dependent upon federal or state forests or the refuse from the wood manufacturing companies. Outfits like Crown and International Paper have their own tree farms, their own foresters, and their own loggers. Without hard-headed planning, Crown would be out of business in a decade at the rate it uses timber.

The CZ company manages 770,000 acres of timber land in the Pacific Northwest—516,000 in Oregon, 254,000 in Washington—and also has holdings in British Columbia and Alaska. Six plants in the two states produce 1,138,000 tons of paper and 160,000,000 board feet of lumber annually, and the locales of these plants ring a familiar bell of the decades in the timber industry—Port Townsend, Port Angeles, West Linn, Wauna, Lebanon, and Camas. The combined payroll is $90,000,000, while about $100,000,000 is spent for goods, materials, and services to keep its operation going. Production includes newsprint and magazine stock for the nation's publications, centered in the East; directory stock, lightweight printing publication paper, towels, napkin bags, toilet tissues, bleached kraft paper for bags, corrugated shipping containers, frozen food wrap, bread wrap, waxed bags, pouches, labels, and electrostatic printed papers. Packaging is a big business with Americans, who are hypnotized in the marketplace by the outer coverings of the goods on display.

One CZ plant also produces 400,000 pounds of ink per month. The largest mill in the chain is the historic one at Camas, with an output of 360,000 tons of paper annually on three machines. However, the all-new Wauna mill, a $110,000,000 investment that began production in 1965, symbolizes CZ's all-out movement into the age of moon shots and space station laboratories. Where once stood a rustic sawmill and company town in the highball days of cut-and-get-out, the four electronic and computerized mills produce daily the equivalent of a 10-foot-wide paper ribbon 4,320 miles long, and 280,000 tons of paper annually on four machines. About half the necessary pulp comes from Wauna's own managed forests of this lower Columbia area adjacent to the Tillamook Burn, and the remainder from lumber and plywood plant residuals, which are converted into wood chips and fiber.

Logs from eight to 50 feet long are rafted to the plant along the Columbia in the traditional manner. But there the old ways end. A powerful clawlike crane hoists the logs in jackstraw bundles weighing 45,000 pounds to the wood deck. A second crane works them onto conveyors leading to the wood mill, where rotating knives strip off the bark, and then a high-speed chipping machine turns a log to chips in about 30 seconds. Chips also arrive by water barge and truck, and with the latter, a sophisticated hydraulic lift tips the trucks to 60-degree angles so the load slides quickly out the back end. About 150,000 units of chips and sawdust (2,400 pounds per unit) can be stored and sorted by a blower arrangement that deflects them into separate piles of particular wood species.

Wauna uses two pulping operations: the "kraft" method and the ground wood process. "Kraft" comes from the German word meaning "strength" and is a procedure of "cooking" the wood in an alkaline liquor to separate the wood fibers from their natural glue, called lignin, which binds them together. There are two kraft digesters: one 230 feet tall, which produces 625 tons of pulp daily from chips and a second that turns out 175 tons of pulp daily from sawdust. About 95 per cent of the chemicals in the spent liquor are recovered for reuse.

The ground wood process is aimed primarily at newsprint production, reducing wood to fiber with twin steel discs operating in refiners. This mill's capacity is 450 tons daily. Both systems are highly automated, as is the entire plant, and controlled from special instrument rooms, which have taken over the thinking and guesswork once handled by the workers.

The great masses of pulp fiber flow on water from the storage tanks to the four paper-making machines. The machines turn out 120,000 tons of newsprint annually plus a range of paper goods from envelopes and tablets to that most important item of modern life, computer cards. At the far end of the line rapid packaging equipment covers and wraps the finished paper products at a rate of 3,000 to 6,000 cases a day for shipment around the world. The steady volume of all Northwest paper plants is astounding; were Henry Pittock around today, he wouldn't be concerned about having enough paper to publish his newspaper, which is still in existence.

The high pinnacle of instrumentation of the entire Wauna plant, with all its complicated mechanisms, even extends into the elaborate air and water control systems, areas made sensitive in the Northwest in recent years by environmentalists. The timber industry has been hard pressed to reduce its pollution of rivers and the air, particularly in Oregon, which has set national standards of new ecological legislation to force reduction of all kinds of industrial and people pollution. The kraft mills have been particularly under fire to do something about their rotten-egg stench—and that is indeed what it is—which can be smelled in strength, with some sickening reaction, for many miles if the wind is right. One kraft plant at Albany was described by Oregon's colorful-talking environmentalist governor Tom McCall as a "blight on the green bosom of the Willamette Valley"; and it has been a standing comment for decades among Portlanders that when the wind is blowing downriver along the Columbia, "you can sure smell Camas."

At Wauna, 40,000,000 gallons of Columbia River water are used daily for paper manufacture, then passed through a system of filtering devices, screens, primary-treatment equipment, chlorination, and clarifier to purify the water before returning it to the river. This intensive filter system is aimed at preserving river quality and the aquatic life. Highly sophisticated odor and dust control equipment is also in operation to maintain the air quality around the mill itself. The odor has been reduced but not completely eliminated. The offending chemical, hydrogen sulfide, remains detectable even when mixed with several billion times its own air volume. The stuff is very potent, and that's a

fact! However, company scientists are still striving for further reduction of
the bad smell, and officials contend that the high degree of instrumentation at
Wauna has contributed greatly to lessening pollution because the entire production process is under finer control and there is continuous testing of this
process all along the line. Odor devices attempt to neutralize the stench of
the sulfide compounds by mixing them with air and other chemicals in an
oxidation tower where the gases are "scrubbed" with chlorine and caustic solu-
tions for further deodorization. Another system involves an electrostatic pre-
cipitator, several of which are used to remove chemical dust from flue gas
during the process when solids of the pulping are being recovered and burned
for reuse.

All the big outfits like Crown Zellerbach maintain laboratories that conduct
continuing research into environmental problems like air pollution and also
work closely with research groups, universities, and government laboratories in
hopes of advancing the methods of keeping pollution at a low level. Millions of
dollars have been spent, and more will likely be spent converting wood-processing plants to methods that will contribute to a "better environment." The
costs are tremendous; early in 1973 Crown Zellerbach announced it would
spend $14,000,000 to reduce the fragrance of its Camas plant. Being practical,
the timbermen realize that in an age of environmental sensitivity, their future
depends not only upon how to grow better trees and harvest them in less
wasteful ways, but also on being able to satisfy a public that is growing more
sensitive and more demanding all the time about the sights, sounds, and
smells throughout the Big Woods.

Chapter XVI

War of the Environment

If they don't stop turning this state into a park, we'll all be out of business. . . .

Remark of industrial forester,
made to author at dedication,
Western Forestry Center, 1971

A few years ago Bill Lyda, who along with his father was blamed for the great Tillamook disaster, went back to Forest Grove to attend a funeral. It had been years since he had been in his home area, where public attitude had made it unpleasant to live. Outside the chapel following the ceremony, an old-time lumberjack confronted Lyda and draped a heavy arm around his shoulders.

"Bill," he declared, "someone oughta erect a monument up at Consolidated Camp to you and your dad. We were all down and out in 1933, and you gave us more damned work than we ever heard of."

It was said partly in logger's jest, but for the place and the Great Depression years, it was true enough. That was the attitude with a great portion of the logging and lumbering community. The forests were expendable. As if there weren't enough jobs after the Tillamook exploded, someone set the Wolf Creek fire in the Oregon-American holdings to the north, which called out additional crews of fire fighters. That was arson, for sure, and there is also the mystery of the second Tillamook fire, to the west of the Lydas' outbreak. It might, too, have been arson. In the starvation-ridden thirties, fires were often set to create jobs.

"You have to remember," stressed Lynn F. Cronnemiller, who was state forester at the time, "that nobody looked on trees in those days as being of much value. And that was particularly true during the Depression, when they were practically worthless."

The burning of empires didn't outwardly bother the old-time lumberjacks. They accepted it as they did sudden death, matter-of-factly. As one told me,

"It was just something that happened." To burn a forest wasn't necessarily any great tragedy. Not everyone felt that way, but a lot of people did. Attitudes have altered drastically toward the importance of timber as a cornerstone resource in the near half century since the Tillamook, especially in the area of carelessness with fire. The American viewpoint has changed very much, particularly since the growth of the ecology movement of the 1960s and 1970s.

"People are a lot different today," says Ed Schroeder, the present Oregon state forester, with some optimism in considering the future safety of the new Tillamook Burn Forest.

The old-time lumber leaders who joined with Colonel Greeley in recognizing that timber was a crop knew what the colonel knew: It would never work so long as fire was a major factor in destruction of the Big Woods. But in the 1970s the winds have shifted; disease and insects in epidemic proportions now constitute a critical and destructive factor in the forests, and with the banning of the use of DDT, even in emergency situations, practical foresters are at loggerheads with environmentalists over the best methods of preservation of disease-ravaged woodlands. In the early 1970s the tussock moth destroyed more timber in acreage and board feet than in the Tillamook disaster, creating a fire potential that would dwarf the Tillamook. The success of salvage remains a moot question, yet it is a point of irony, for if fire destroyed that much timber, especially if originating from a logging operation, environmentalists would be storming the governor's office and picketing the state forestry headquarters. Environmentalists advocate allowing nature to take its course, no matter what the loss. It is also their view that the demand for more and more lumber for American and foreign consumption brings very heavy pressure for extensive cutting on public lands (in 1973 the Nixon administration raised the quotas for harvest in federal forests) and that this is a far greater threat than destruction by fire and disease. All this is linked directly to the war between the timber interests and the conservationists in what has become the greatest movement concerned with the environment in the nation's history. And the loud-raging debate among ecologists, timber people, and public foresters has many facets, from the clear-cutting of forest tracts to the establishment of parks and hiking trails, with no end in sight.

The forests are as flammable as ever, especially in a long, dry year, but the aforementioned change of attitude, coupled with hard-hitting laws against itinerant logging operators; the social pressures put upon the careless by sportsmen's clubs; the tender, loving care a large portion of the public is now giving its outdoors (other than litterbugging the landscape); and far more effective fire-fighting methods have brought about this generally fire-safe condition in the forests. One of the Big Woods' greatest fire dangers today comes from out-of-state summer tourists of big urban centers such as San Francisco, Los Angeles, and the East Coast who haven't been educated to using caution with cigarettes and campfires. Thus the Keep Green movement in the Pacific Northwest remains alive and well as a continuing reminder to these part-time outdoorsmen with signs, radio and television announcements, and a variety of gimmicks to catch the public's attention. And it continues to pay off. Oregon, for example, had a great many more fires in 1972 (3,238) than in the previous year (2,280). Of these, 1,737 were man-caused in 1972. But the acreage was almost half—7,592 acres burned by men and lightning, against

13,657 acres the year before. And, of course, there is no accurate way of telling how many more fires might have broken out without the Keep Green activity.

Just how drastically things have changed since the "good old days" was exemplified in the state's Southwest Oregon District, where 1,800,000 acres of state, private, and Bureau of Land Management lands are under the supervision of Curtis Nesheim, the rugged fire warden of the Northwest Oregon District during the early years of the Tillamook reforestation program. In 1972, 212 fires were reported: 57 from lightning and 155 attributed to man's activities. *But none was caused from logging or other forest industry activity!* Also, in the Rogue River National Forest, which had 124 reportable fires, only one was charged to logging when a rubber-tired skidder overturned and burned. It demonstrates a progress that is yet difficult to believe for anyone who remembers when the summer skies were always ablaze from cigarettes, warming fires, and logging in low-humidity weather by lumberjacks who couldn't care less.

When fire breaks out today, crews hit it hard and fast, as if attacking in an all-out war. There is no holding back. Aerial patrols, networks of access roads, radio communications, modern, powerful equipment, and chemical warfare enable protection crews to reach the front lines in a matter of moments from the fire outbreak. Accessibility of all the forests is the key to bringing a remote fire under almost immediate control. Smoke jumpers, who were pioneered years ago to plunge into the rugged back country of Montana, are now commonplace, trained in U. S. Forest Service schools at Redmond and Cave Junction, Oregon, and Winthrop, Washington, to parachute directly into the woods where the flames are raging. Helicopters land crews in the wilderness in a hurry, while the road systems, which cost the timber industry and state and federal forest budgets as high as $75,000 a mile, pay dividends by moving heavy equipment quickly to the fire lines. And the forest districts have managed to build up a wide array of equipment far different from the shovel and ax, which were standard in yesteryear. If roads lead into the fire area, there's a land invasion of heavy rigs—500-gallon tank trucks, plows, 'dozers, trench diggers, brush cutters, pumps, power saws, and other gear, so that the stampeding flames can be turned and contained. The fighters wear protective clothing, which lessens the dangers of being burned by flying brands and sparks. And from the air, patrols keep constant tab on the fire's progress, intensity, change in direction, and leaps into new areas, while heavy DC-6 "bombers" drop tons of fire retardant chemicals to knock down spot fires.

Some of the worst fires of recent years have been in Alaska, where raw old growth is being harvested and where forests haven't yet come under the intensive management system now so prevalent in Washington and Oregon. Foresters and protection crews are still dealing with wilderness back country where logging hasn't reached, but lightning knows no bounds.

In the land of the Big Woods, the investment in new forests and the mounting importance of timber as a national resource for the future make fire carelessness intolerable. Slash burning as a common practice is obsolete, for private timber interests have found the value of wood waste too significant to go up in smoke. There's also the matter of air pollution, and in the Pacific Northwest these days watchdog environmentalists are behind every tree and rock. What remains of a harvested area that can't be removed for economic gain is allowed to return to the earth to feed new life into another generation of timber. Controlled slash burning is still practiced, however, in isolated cases, particularly in national forests where steep mountain slopes and lack of good roads make it impossible to bring in equipment, let alone operate it, to chop up debris into mash for returning to the soil, and where it is unfeasible economically to haul it out.

When it was suggested in 1972 that perhaps an occasional forest fire might be good for the woods, foresters who had long battled public carelessness with education were horrified. It brought a quick angry response from Ed Loners of Keep Washington Green and other timber spokesmen, since the public was being given the wrong slant by ambitious scientific researchers who were advocating fire to help maintain a balance in the ecosystem and who were spreading the word in dramatic headlines that perhaps *"Smokey the Bear has been lying for years"* and that fire prevention *". . . is grade school stuff, and grade school is where Smokey belongs."* This "advanced" modern theory was threatening to tear down all that had been built up over three decades by the Keep Green and Smokey the Bear movements; the human animal being what he is, it wouldn't take much to encourage him to backslide into fire carelessness.

"Statements like these puzzle and infuriate us," Loners roared. "They confuse the public and, some become apathetic to forest fire prevention. We don't deny the intelligent scientific statements of these men, among the foremost perhaps in the field of fire and fuel research and management; but we certainly do take issue with how their new knowledge and wisdom are ladled out to the public. The facts and expected results are not put out in balance for public consumption. Some of it is damaging to our Keep Green forest fire prevention program and objectives."

Thus the matter of fire continues to be a prime concern. Fire kills seeds and seedlings just beginning to gain a roothold in their search for the soggy sky, and also the litter vital to the formation of humus. It may alter completely the complexion of a forest by wiping out existing species and paving the way for faster-growing but less valuable tree varieties. One of the big problems of the Tillamook Burn in recent years has been the growth of alder, which threatens to squeeze out young Douglas fir seedlings before they can become tall enough to hold their own. As a result, foresters under the direction of Bill Hoskins, reforestation forester with the Northwest Oregon District, have been engaged in an intensive scarification program to eliminate the alder and open up the land to the small conifers. Another impact of fire is that it clears the way for disease and insects, which kill and ruin even more trees. The fires weaken the trees that aren't destroyed, making them vulnerable to such infestation. Industrial foresters today search for infected trees with the same vigor applied to locating an abandoned campfire, or lightning strike long smoldering in a snag. The bark beetle, parasites, the tussock moth, insects that attack cones and seed—these are the big threats to the forests of tomorrow. Deer, bear, and porcupine take their tolls of young timber; and once again there is the problem of keeping things in perspective between environmentalists who oppose the killing of wildlife, and the practical forester trying to save the timber for the future, as to just where the values lie and in what degree of balance. The woods are as filled with dangers for the survival of trees as the seas are dangerous for fish. And under the ecology movement of the seventies, restrictions in the use of chemicals, most of all DDT, have given foresters much cause for concern. In 1972–73 environmentalists were faced with a heart-wrenching decision because of their intense opposition to the use of DDT against a critical tussock moth epidemic in eastern Oregon and southeastern Washington. Foresters maintained that only DDT would halt the destruction by the moth in a huge 500,000-acre region, and the likely loss of the timber as salvage lumber also. Environmentalists were being forced into a position of yielding "if there is no other way." Yet the federal Environmental Protection Agency, despite appeals and pressure from even the Oregon governor, banned the use of the dread yet effective chemical in the hope that through natural cycles and use of acceptable yet less effective and short-lived insecticides such as Zectran, the moth epidemic would burn itself out. Again, it was a matter of priorities, although suspicious environmentalists charged that foresters were at-tempting to break down the DDT ban in this manner, and its use on the tussock moth would open the door for additional application elsewhere in the future. Yet the moth was leaving a dead and dying forest, highly flammable, in its wake, a waste of millions of board feet of good, sound timber at a time when the demand for lumber was at an all-time high and critics were making much over the volume being cut in federal, state, and private forests. Zectran didn't work; the moth continued its destruction. So the following year, after extensive damage had been done during these years of delay, the way was cleared for spraying that area with DDT, which put an end to the moth's dev-astation. But the debate goes on, and the broader question of just which route is the best to take in moving into the future remains unresolved.

The forests of the future represent a tremendous investment, averaging about $30 a tree, not only on public lands but also in the immense holdings of the private timber companies, which they like to call "working forests." Weyer-haeuser, the sprawling Big Daddy of the industry in the Pacific Northwest, with holdings totaling some 2,700,000 acres in Oregon and Washington, in 1967 be-gan an intensive program of wide-scale tree farming that it called "high-yield for-estry." It falls under a variety of names, from "forests for the future" to "multi-ple-use forests" and "forests for a better environment"; these great-sounding monikers are often the products of the wordsmiths of timber advertising and pro-motion departments devoted to telling the story to the public in an appealing but still rather honest fashion. The aim of these Space Age tree farm projects, tar-geted into the next century, is the continuing production of timber under a grandiose rotating system of intensified management for "growing trees faster and better than nature can do unassisted." Or, put more simply, "growing more wood in less time." It is expanding and improving on the early tree farms and also upon the hundred-year cycle pioneered a quarter century before by the Simpson Timber Company on the Olympic Peninsula. The cycle has now been amazingly halved, from 80 to 100 years down to 45 to 60 years. Industrial foresters have be-come scientific farmers or gardeners in the broadest sense, geared to "handle a tree crop like a garden." The tree farm movement that Weyerhaeuser and the in-dustry's Joint Committee on Forest Conservation helped launch in 1941 has blossomed far beyond the wildest dreams of its advocates; much of this has come about not only through the control of fire and disease, but also through amazing discoveries in quiet forest laboratories and through the use of computers to quickly select and chart the course for growing better trees faster in a variety of timber areas, each with its own special conditions of soil, climate, and altitude.

The timber companies maintain that they can keep pace now with the annual harvest in new forests. The "ripe" stands are harvested and replanted as soon as is feasible after the land has been prepared or tilled for the new seedlings. Helicopters spread chemical fertilizers and may even do some aerial seeding, although hand planting remains the most effective method. But even that has been speeded into the Jet Age. As the forest grows, it is thinned precommercially of the weaker seedlings to give the stronger ones a better chance. Weyerhaeuser grows some 150,000,000 seedlings a year in its nurseries, and more and more, these are coming from what are called "supertrees." Growing isn't left to chance, as in the parks or natural forest areas. In 15 to 20 years the forest will undergo a commercial thinning, the wood from this operation usable as chips and pulp. The remaining stand can then spread out to its full development for a complete harvest, probably by clear-cutting, in another two or three decades, when the trees, at least by industrial standards, have reached their ripened maturity.

Other great companies such as Crown Zellerbach, Georgia-Pacific, International Paper, Boise Cascade, and Pope & Talbot are involved in similar trees-for-the-future programs in their holdings. It's a matter of survival for the companies, which are now bigger and more permanent than the men who run them. The industry, state forestry departments, and the U. S. Forest Service have combined efforts, through their forest laboratories, experimental nurseries, and tree farms, and work for better forest protection and production by swapping ideas and trading data and information for the general welfare of the Big Woods.

Unfortunately, things are changing so rapidly in the mechanized Big Woods that at times the U. S. Forest Service finds itself lagging behind private industry in the adoption of new ideas, methods, tools, and equipment, even though its competent foresters are certainly aware of them. A lagging Congress is to blame by failing to appropriate the proper sums for advanced Space Age forestry as being practiced on private lands. Time and again the people in Washington, D.C., have failed to understand the needs and the problems of the Big Woods. As an example, a new project called, in the fashion of the times, FALCON (Forestry, Advanced Logging, and Conservation), was set up as an experimental program of the federal government in 1972. It was designed to pioneer new logging practices through a congressional appropriation of $5,000,000 and was estimated to cost $10,000,000 over the next decade. The program was forced upon the Forest Service and indirectly upon state foresters and the industry by the environmental movement, the soaring demand for lumber, and the increased dependence upon timber lands that are difficult to manage or are very remote. Another factor was public pressure for reduced timber acreage due to more and more forest areas being designated as untouchable wilderness, parks, and recreational preserves. One third of the FALCON budget went for research into environmental matters, while the remainder was being used for experimentation with aerial logging. Then the funds were cut off, but the Forest Service saw fit to carry on certain phases of FALCON. Through this Forest Service encouragement, 'copter logging advanced by 1973 beyond the experimental stage, although considered by cautious foresters to be still in a "stage of development." But the need for more timber was reaching the critical level. While logging on federal lands is limited by law, sound trees exist in the back country that might be harvested if loggers could get to them

70. Use of helicopters in lumbering and forestry has come a long way since early days of Tillamook Burn experiments. Helilogging is the newest means, although expensive, for getting logs from environmentally sensitive areas by lofting them above the treetops. Split-second timing is necessary in helilogging, for every second costs many dollars in the three-minute trip from woods to landing. The logger at left is rolling up chokers for return to the woods. Loggers don't wear traditional red hats any more. —*U. S. Forest Service*. (Chapter XV)

71. Balloon logging is another method used by some timber companies to get out the big sticks under environmental restrictions. —*U. S. Forest Service.* (Chapter XV)

72. Clear-cutting, or patch-cutting, is an accepted logging practice in the Big Woods, although it has come under much fire in recent years from environmentalists. Foresters say the Douglas fir grows far more rapidly under sunny conditions, returning the area to productivity sooner. —*U. S. Forest Service*. (Chapter XVI)

73. In modern industrial forestry, timber is given tender, loving care, even fertilization from the air. —*Georgia-Pacific*. (Chapter XVI)

74. Forest nurseries and tree farms had their beginnings in the Big Woods. Industrial and public nurseries are found throughout the region, producing millions of seedlings annually. This Weyerhaeuser Little Rock nursery in Washington has hit a peak inventory of 59,000,000 Douglas fir seedlings. —*Weyerhaeuser Company.* (Chapter XVI)

75. In developing supertrees and seed orchards, foresters hand-pollinate special trees in the woods to get the best possible seed. —*Weyerhaeuser Company.* (Chapter XVI)

76. Superseedlings are now grown in huge greenhouses by the millions in "plugs," so fast that they are ready to set out within a few months. —*Weyerhaeuser Company.* (Chapter XVI)

77. By growing seedlings in plastic cells or plugs, roots are better protected and undisturbed. Being planted in its own protective soil, the shock of transplanting is less, and seedlings take hold right away in woods. —*Weyerhaeuser Company.* (Chapter XVI)

78. In 1973, 40 years after the great Tillamook fire, foresters, loggers, lumbermen, public officials, and citizens gathered at the old Owl Camp to hear Oregon governor Tom McCall dedicate the new Tillamook Forest, against a background of new timber. State forester Ed Schroeder is on platform at right. —*Ellis Lucia photo.* (Chapter XVII)

79. During postdedication tour, people saw at close range how the Burn has been cleaned up and the charred landmarks are being hidden by the new growth. —*Ellis Lucia photo.* (Chapter XVII)

80. The Tillamook Burn, where modern forestry had its beginnings, is green again, with the replanting of mile after mile of rugged terrain. Some 73,000,000 trees in all were hand-planted under a 25-year public program. —*Ellis Lucia photo*. (Chapter XVII)

without damage to the land. This is where the helicopter, even more than the balloon logging system, is coming on strong, not only for remote standing timber, but also for the salvage of windthrow and firekill.

New ways to harvest timber with less damage to the land . . . new methods of regrowing the forests for the next and the next generations . . . new ideas for utilizing every last bit of the tree . . . these things occupy much of the time and thinking of private and public timber people today, now that everything has become so fully mechanized. The development of the so-called supertree promises a quicker turnover by growing far more rapidly a healthier, sounder, taller, thicker, and straighter tree than the Douglas firs of the past, or any other tree type, for that matter. These trees will not only yield more wood, but also be more resistant to disease, insects, and fire. Some environmentalists charge that the foresters are trodding on dangerous ground by producing a highly bred tree that won't be able to grow without the tender, loving care of mankind and that, if at some future time man's decline is such that this care can't be given, the forests will die. Industrial foresters chuckle at this argument and retort that the environmentalists voicing this viewpoint don't know about what they are talking. The supertree is nothing so special, they say. It is merely *good breeding,* whereby through the application of the principles of genetics, the best . . . the healthiest . . . the finest . . . the tallest trees of a forest are being turned into "parents"— the forebears of generations of fine timber for the future.

In the northwestern corner of Oregon, adjacent to the Tillamook Burn, the Crown Zellerbach Corporation is involved in a typical experimental program with the Oregon Forestry Department, International Paper, and Longview Fibre Company. Development of a special tree, like a thoroughbred racehorse or a good line of cattle, is a long and tedious process that early-day loggers would never understand, although you can't help but reflect upon the legend of the Douglas firs on the Humptulips. Foresters like to call this a "genetics improvement program." It is of no short duration!

"A lot of our effort, when you're talking about a supertree, is nothing more than breeding a better cow," explains Robert Lindsay, veteran forester with Crown Zellerbach, whose career stretches from his high school days when he wrote themes about forestry. "It's one step at a time. You take a good bull and some good cows, and you breed; and you take the best calves and you breed them, and eventually you breed a better herd. Of course, with a herd of cows, this is easier to do, a shorter-term stand than with a forest. A forest is far more difficult to breed."

In the beginning, 1966–68, foresters scouted the region, looking for the best trees to become "parents." Such trees are rated "better than average," and as Lindsay describes the approach, "it all depends on what you are looking for" in terms of future generations of trees.

"A lot of judgment goes into this, and you have to draw some guidelines," he adds. "We're not sure what we are looking for; everybody has his own ideas."

Among other things, foresters judge a tree by its height and diameter growth, the way the limbs branch, and how many limbs are in each whorl. The simplest method is to secure cones from these parent trees. In this co-operative project, 900 trees were selected and the cones gathered. The seed was thrown into a central pool, the foresters designating it as "better seed for use in the northwest area."

Then foresters advanced the project several steps. There were signs in January

1971 of a big cone crop, and since the Douglas fir bears cones only every five to seven years, the way was opened for the controlled breeding of future trees, through hand pollination. Artificial insemination over natural sex had come to the forests. But this is more difficult than it appears, unlike pollinating flowers in a nursery, since these plants are 70 to 80 feet tall.

First, the foresters selected the firs to become the male trees and then those to be the female trees, since each tree contains the qualities of both sexes in its flowers. The marked female trees were climbed in March and plastic bags placed around limbs showing cone buds to protect them from contamination by pollination from substandard trees. It was a tough, tedious job, for March weather in the Coast Range isn't balmy as winter winds down. But about 160 trees were climbed and bagged, at a rate of four a day.

The next month, the crew returned to the woods to scale the male trees, each bearing an identifying number. The pollen was collected, placed in vials with corresponding numbers, and then dried. In May or June the pollen was returned to the same woods, the female trees climbed once again, and the pollen injected into the bags with a syringe, for the female cones had now come into flower.

Finally, the following September, foresters went back to harvest the cones, bags and all, taking them to their laboratories for processing the seed. Throughout the entire process, extensive records were kept so that foresters may quickly determine the parents of any subsequent seedlings, which would number eventually in the millions. Normally, foresters would be working with "wild seed," whereby they might know the mother but not the father. But in this particular reproduction project, both parents were known.

The next spring, a year after first working with the parent trees, the seed was planted for the beginnings of a seed orchard. Yet it will be 15 years before the young trees are of cone-bearing age, and the investment in each will be from $15 to $30.

"So you have to have faith," added Lindsay. "Patience and faith."

The investment is comparable to a farmer planting a fruit orchard and bringing it to bearing age, he points out. But this 30-acre seed orchard will produce all the seed needed—and of the finest quality—for the 350,000 acres in that area owned by the four participants. And on a 60-year cycle, harvesting would be only 6,000 acres a year. Consequently, the effort is considered a wise investment in the future, since the need will be only enough seedlings to replant those 6,000 acres.

"Although it's a long program," Lindsay concluded, "you've got to do something so that 10 years from now, you don't look back and say, 'I wish I'd done something.' We want better seed, and it will be 15 years before we get it."

High labor costs and the urge to progress constantly faster have spurred foresters and forest scientists into growing better trees more swiftly. But seedlings are tender plants to be handled very gently, with the sensitive roots protected from light, bruising, and lack of moisture; and since they are expensive, along with the soaring wages being paid planting crews, the necessity of logger-style inventiveness again enters the picture. Hand planting has proved consistently more effective than aerial seeding, with a much higher survival rate; therefore, foresters have been seeking better methods than the old stoop-and-bend-grunt-and-groan, using the traditional hoedags as planting tools. Also, there was the problem of shipping and handling the tender seedlings, always with the danger that the roots would become overly dry and hot, thus delaying growth when set into the ground or perhaps never taking hold at all. In the bare root system, used by foresters for generations, a lot of seedlings died from air exposure and lack of care.

Planting guns were being developed in the early 1970s that would virtually "shoot" the seedlings into the ground. Canadian foresters in British Columbia devised a method using a tiny plastic ball or bullet surrounding the roots, then dropping the seedling quickly into a hole in the soil with a small "gun" or tubelike contraption. But the Canadians were having trouble with the bullets because plastic won't decompose or break down, so the roots were kept locked up, and growth was delayed or stunted. Later, the Canadians devised a method of splitting the bullet as it dropped, releasing the roots more quickly.

Phil Hahn, a Hungarian forester with Georgia-Pacific, took the Canadian idea another step. Why not grow the seedlings in small blocks of bullet-shaped dirt? Plastic and Styrofoam containers were designed with cone-shaped holes about five inches deep and tapering to a point on the lower end. Good, rich soil was used and the seedlings grown very close together, with about 100 seedlings to each container. Looking across a greenhouse bank of them is like viewing a very thick lawn of tiny, high-quality trees growing five times faster than do natural plantings. For his planting tool, Hahn devised a "double-barreled shotgun." One barrel contains a dibble that punches a hole in the ground the same shape as that of the tapered hole where the seedling is growing. The dibble is pushed into the soil with a foot pedal, then pulls up from the hole automatically as the other barrel swings into position. The cone-shaped roots, encased in dirt, are lifted from the Styrofoam container and dropped directly and swiftly into the hole. The planter takes a swipe at the soil by stepping near the seedling with his heel, packing it solidly around the baby tree, and then walks another 10 or 12 feet to begin the process again. The only difficulties, it was later learned, were that the rapidly developing Douglas fir seedlings were often too large to clear the tube, and the muddy conditions of Northwest soil during winter planting seasons sometimes clogged the guns. Foresters set aside the tubes until there could be more experimentation to adapt the gun to Pacific Northwest conditions, but they continued using the dibble which, in itself, speeds the planting process about double that of the old hoedag system, to around 1,000 trees per planter a day. However, the full gun is being used in other parts of the United States, primarily in the South.

Georgia-Pacific foresters claim they are getting 95 to 99 per cent survival, compared with 60 per cent survival under the old system, since the supertree seedlings are very healthy and growing, not stagnated, the roots aren't bound or dry, and they aren't hampered by a plastic encasement or anything else to slow down the growth. Georgia-Pacific took out the patent and manufactures the new tool, making it available to other companies and forestry departments. One rival company quickly placed an initial order for 200. As has happened so often in the Big Woods, the new system is revolutionizing tree nurseries, for seedlings are already being grown in greenhouses by the many millions in cone-shaped containers that are filled with soil and planted by mechanization, assembly-line fashion. Had this all been developed two decades earlier, the laborious task of replanting the great Tillamook Burn might have been accomplished in half the time, other things being equal.

The greenhouse seedlings are far ahead of those grown in the outdoor nurseries, for the cone-shaped roots are solid and alive rather than dry and perhaps rootbound, as nursery stock may be by the time it's set out after two years or so. Using the new supertree stock, the fast-growing greenhouse seedlings are only nine months old when planted, the eager roots spreading soon after they hit the soil. Foresters estimate that they have a two-to-three-year head start on the nursery stock in the growing of a new forest and a far better chance of survival. The mortality rate on nursery stock is sometimes very high.

Space Age foresters, more than ever before, are combining scientific advancement and technology with what Nature already provides. As the developers of big rigs aim to help the logger, the object in modern forestry is to "enhance Nature" rather than oppose her, buck her, fight her, or attempt to change her. There are wonderful natural-growing qualities in trees, especially the Douglas fir, and foresters of the Big Woods are well aware of them. Nature might certainly regrow a forest, as she has done in the Little River Valley of northern California's redwood country, or in many parts of the Pacific Northwest, but in Little River it has taken half a century to do so. The broadcasting of seed on the winds of the gods from the high ridgetops where mature trees are left standing certainly works, and for its time it was the best system the lumbermen and lawmakers could devise. But this is a long, slow, hit-and-miss process, which isn't acceptable to the commercial forests of modern America if the woods are to survive and not be depleted in the manner of other great natural reserves such as oil, gas, and minerals that are beginning to run frighteningly low. Foresters are trying to help Nature do a better job by sidestepping the gambles of hit-and-miss.

But in backing Nature's plan, industrial foresters are committed to the policy of logging off the mature timber on a year-by-year basis, and this is where public and private foresters and the timber companies run head on into a clash with the environmentalists, who quickly point out that Nature has no loggers and that in Nature, the forests are allowed to live as long as they will and then die, to be replaced by other species of trees.

Not true, argue the foresters. Nature burns the woods, setting them afire with lightning strikes. She "logs" also through violent windstorms such as the great Columbus Day blow of 1962, which downed billions of board feet of prime timber; and she brings in disease and insects with their own destruction, another method of natural "logging." The old, dead, and dying trees are extremely vulnerable to lightning strikes, and (point out the foresters) the resulting fires destroy more than just the dead wood. New and young growth is also consumed. Not to harvest this mature growth—the term "log" is being carefully eliminated from the industrial forestry vocabulary because of its past connotations in the public mind —is pure waste from the timbermen's view.

Environmentalists retort that so be it; let Nature decide what is right and man leave her alone, accepting his own place in Nature's plan, rather than trying to remake her in his image. This is the philosophy and approach of the national park system, where nothing is disturbed beyond providing for the safety and enjoyment of park visitors. The forests, the glades, the woodlands, all else are allowed to grow and die or change in form without tampering, even though the temptation is there. Many timber leaders would like to log in the national parks, since they see waste there, by their own standards and points of view. Sound timber will otherwise rot away and thus be wasted for man's use, other than for its genuine aesthetic value, which of course ranks very high with the public. The untouched scene in Nature's own handiwork is something of beauty and inspiration that needs to be preserved as it has unfolded and not as some men would like to have it. Some bits of our world deserve to be left alone, and the national parks are among these places.

"The public is going to get a shock some summer when much of Yellowstone Park catches fire," observed one timberman. "Much of the timber there is dying; there are thousands of dead and diseased trees. Yet nothing is being done to save that forest."

The environmentalists say "hands off," and I agree with them. If Nature wishes to alter Yellowstone, even turn off Old Faithful, we should allow her to do it.

In the 1970s the timber industry and state and federal foresters of the Big Woods are having to shape their policies and techniques to the greatest public outcry over what is called "the environment" in the nation's history. Conservation movements have blossomed before, but nothing to the extent or toughness of fiber of the current movement; spearheaded by the nation's young people, it is receiving the strongest kind of support from the over-thirty population of middle-aged and elderly. Since the mid-1960s, when young people were casting about for causes outside the immediate political realm, the movement has gained rapid momentum, which stunned many officials of business, industry, and politics, especially regarding the demands being made upon them to quit using the air, rivers, lakes, and land as garbage dumping grounds. The momentum has been especially strong in the West, where people have a particular love for the land and the scenic places that are part of their heritage, and who have when traveling in the East seen what, with utter disregard, the great industrial centers have done to land, water, and sky, and in Los Angeles, where the forests are dying from smog poisoning.

The environmental movement held forceful significance for the Pacific Northwest, especially Oregon which had long maintained something of a status quo between the good, easy-paced life and rapid growth and development. This was a condition that was most enjoyable to its citizens, who loved their seashores, mountains, plains, rivers, lakes, and forests with a passion, all of which were rated as not only very beautiful but also among the cleanest and most pollution-free in the world. Their point of view was a direct outgrowth, or by-product, of the Big Woods. Much impetus and many ideas and leadership for the national ecology movement have emerged from these woods. The first national Earth Day chairman hailed from Camas, Washington, the paper mill town. And Oregon became a front-runner and testing ground through its ecologically sensitive people and a governor who has been outspoken in his views about protecting the natural wonders of his state and the nation. As a result, no industry felt the impact of the movement greater than has the home-based wood products manufacturers.

Oregon growth since wagon train times had been slow and easy, save for occasional spurts such as those following the 1905 Lewis and Clark World's Fair, and during and immediately after World War II, when thousands flocked to Portland to work in the shipyards. Most of these gradually left, largely from a lack of good jobs and high wages (Oregon's pay scale remains well below the national average), but there was a certain spinoff that increased the population. Timber and its allied firms remained the leading industry, with agriculture a strong second. There was little encouragement to solicit big new industries or to woo fat federal government contracts. Oregon politicos didn't have the clout, and the citizenry, other than ambitious civic leaders in Portland, who claimed they wanted to build "another Los Angeles," were perfectly satisfied to allow things to continue along the lines of the past, their families and friends enjoying all this rich, lush beauty and clear, trout-choked streams without having to fight for space, as in a huge urban society. Later, they were accused of trying to keep Oregon for themselves, which was by and large the case. On either side in Washington and California were prime examples of what happened when you sold your soul to bigness and dollars. The basic values of those two states, and the resulting changes that were wrought, were far different from the unique conditions that existed in Oregon.

Yet the core of the Big Woods made a prime mistake, although it seemed like a good idea, in proudly promoting its beauty spots for tourism through state-financed national advertising. Tourism grew into the third-largest industry and was believed to be a "clean" one. Suddenly many eyes began turning to Oregon, which had been largely bypassed because of its wet climate, especially when compared to balmy California; and developers with millions to invest in damned-if-I-care fashion showed no concern for Oregon's many intensely beautiful outdoor places, so long as they got a maximum return on their investment. There were few laws to restrict what they did to the land or places of natural wonderment. Where Oregonians had managed to gain control over unbridled elements of the timber industry, this was a whole new ball game.

In the mid-1960s the 400-mile Oregon seashore, which ranks as one of the world's great coastlines and is also one of the few in the United States to which the public has free and unrestricted access, became the first battleground for preservation of the environment, and eventually caused Oregon to become a national pacesetter in laws and regulations to clean up America, just as the Keep Green movement had done 20 years earlier. Suddenly out-of-state developers began taking much interest in this rugged coastline. They had figured out a way to capture it, despite the public preservation established by generations of heritage and the vision, half a century before, of an Oregon governor, Oswald West, who officially proclaimed the beaches a "highway" to be retained in perpetuity for succeeding generations of Oregonians.

Oregonians were horrified to learn that this might all be broken down, that their wonderful coastal playland might be fenced and barricaded for development and restricted access. They envisioned lines of high-rises such as at Miami Beach and Honolulu. California investors were erecting a motel monstrosity down the side of a beautiful bluff on the central coast. And at Cannon Beach, the owner of a newly built motel right on the sandy strip put up a barricade around "his beach" for restricted use by the motel's own guests. Oregonians were reminded of the scene at Malibu. Still other developers were bulldozing the beaches, changing their complexion, and ruining clam beds. And right there, the top blew off of everything.

Angry Oregonians marched on their 1967 legislature in Salem and in one of the stormiest sessions in the state's history, rammed through legislation that would secure the beaches and coastline for the public for all future time. Developers, including the motel owner at Cannon Beach, tried to break the legislation in the courts, but to no avail, and in the end the barricades came down. But this was only the beginning. The people learned what they could do, and in succeeding legislative sessions have passed tough laws to force a massive cleaning up of rivers, tidelands, lakes, harbors, the water, and the air. Many of these laws, like the so-called Beach Bill, have been used since as patterns for similar laws in other states and bills in the U. S. Congress.

Fortunately, too, Oregonians had a sympathetic governor in Tom McCall, a native who was raised on a central Oregon cattle ranch and who didn't much like what was happening to his state from this sudden flood of investors, tourists, and newcoming permanent settlers who had suddenly decided that places like Southern California were intolerable. Some of the legislation was downright inventive. The residents of the state where the word "litterbug" was coined were heartsick over the junk that was scattered over their beautiful woodlands and beaches, and thrown into streams and lakes. As once Oregonians had grown weary of seeing their beautiful forests destroyed by fire, they were now fed up with this trash problem. A program for picking up litter along the highways and roads was financed through special ego-type personal license plates, costing $25 annually, and this helped. But one of the big litter problems was caused by pop and beer cans and bottles. In 1971, following some intense infighting, the nation's first legislation was passed forcing manufacturers and merchandisers to set up a system for returnable bottles, for a few cents' cash apiece, and also outlawing pulltab cans because the metal rings were doing extreme damage to birds and small wild things. Again, the angry manufacturers, with much financial support from around the country, forced a public vote over the bottle bill and then took it to court. In the 1973 legislature (Oregon's body meets only every other year) they were still trying to break it down, but by that time the law was in force and was working astoundingly well in the massive cleaning up of the state's outdoors. As a result, a national bill was introduced in Congress.

The environmental war in Oregon has been the most intensive in the nation, due to the great pressure for protective laws against those who care only for exploitation. It is of prime importance to a great majority of Oregonians, most of whom consider themselves environmentalists and wish to protect their world at almost any cost. The timber industry, vulnerable and sensitive, was caught in the middle of the preservationists' web and at once became a prime target. A lot of the timbermen didn't like it, because the projected costs would be tremendous, although in its ranks the industry probably has more sportsmen and outdoor types than any other single group.

"If they don't stop turning this state into a park," moaned one timberman at the dedication of the new Western Forestry Center in 1971, "we'll all be out of business."

Oddly, a strange yet effective method of calling attention to population growth and what was happening to the Oregon Country had its roots deep within the timber industry itself, through the man who had long been logging and lumbering's Boswell. Stewart Holbrook had written extensively for international audiences about the Big Woods and its people. Long before anyone else, this lumberjack author had seen what might well happen to the beauties of the Pacific Northwest and with characteristic humor and satire had set in motion the wheels to call attention to the dangers. It began in 1940 with the birth of the Keep Green movement in Seattle.

The movement's backers wished to ballyhoo the new director and the newly declared war on fire as loudly as possible. A press conference and banquet were held to hail Stewart Holbrook's return to the tall and uncut, and to give Keep Green a rousing sendoff. By an odd twist, the event also saw the beginnings of

Holbrook's own private conservation movement which, 10 years after Holbrook's death in 1964, became a significant factor in Oregon's save-the-environment efforts, even to expanding the fire campaign's original slogan to Keep Oregon Green, Clean—and Lean.

Since 1940 was an election year, a young reporter at the press conference asked an off-trail question far from the subject at hand. Perhaps, as Holbrook had just returned from the East, he knew something that hadn't reached the far corner of the Pacific Northwest about who might run for the nation's top office.

"For whom are you going to vote for President?" the newsman asked.

Holbrook, whose mind was fixed in Keep Green, hesitated and was visibly irritated. It took him a moment to gather his thoughts, and he reached for the first name that popped into his head. Half in jest, half in sarcasm, he replied, "Why, James G. Blaine, of course."

Pencils were suddenly busy on notepads. The Seattle newspapers played it straight, with fat headlines, the editors not taking the trouble to learn the identity of James G. Blaine, who they felt must be a dark horse yet to emerge. Stewart was delighted at the success of his leg pulling as part of the Keep Green launching. Later, when the Seattle Advertising Club threw a party to kick off his latest book, someone pinned a huge "Vote for Blaine" campaign button on Holbrook's lapel. The Keep Green chairman found the name of this nineteenth-century politician who almost became President (Stewart always felt Blaine should have been) acceptable and handy, dredged up "in a wild moment" for what became his celebrated James G. Blaine Association. Often when asked to comment on some social, political, or other controversial matter, Holbrook would dodge the issue by answering, "Oh, our James G. Blaine Association will take care of that." His reply seemed to satisfy a lot of people.

The spoofing yet conservation-oriented Blaine Association evolved into an inside regional joke, and characteristically, as his stature as an author grew, logger Holbrook played it to the hilt. Well ahead of his time, in the postwar years of World War II, Holbrook became much concerned over the expanding population of his beloved adopted home, the Pacific Northwest, and what it would do to the forests. His "little association" seemed ideal to combat this growth, through good-humored satire in the fashion of Mark Twain and Will Rogers. Blaine was a Secretary of State who advocated "America for Americans" and who never set foot in Oregon, which was of course the point. Still, in a time when growth was very acceptable and highly touted, Holbrook's one-man "campaign" was distinctly against the tide and the Chamber of Commerce. Even friends weren't certain where the serious historian of a mounting list of nonfiction books left off and the humorist began. As for use of the strange name of someone lost to history, Holbrook replied that it suited his cause.

"It struck me as having the right antique flavor to confuse a public into thinking us a crowd of old-time fuddy-duddies, still living in the past, devoted to some obscure cause, but harmless. Harmless? Harmless like a cobra. We believe that the Pacific Northwest is the Promised Land, and we mean to protect it from the steadily mounting danger of overcrowding by the hordes of Goths and Vandals who have been touring our blessed region in increasing numbers."

Too many people, he warned, would "clutter up the place." A fence around the Northwest wasn't such a bad idea.

"Like many a native," he said, "I am privately of the opinion that not only

Crater Lake and Mount Rainier and the Olympics, and the Oregon Caves and the Lava Beds should be in federal keeping, but perhaps the whole Northwest should be set aside as one great park before it is wholly overrun by foreign immigrants like myself."

Begun tongue in cheek, Stewart discovered by his mail that his supporters were dead serious. He worked hard at his "campaign" as one of several satirical putons he developed as an author. He claimed he was discouraging hundreds, if not thousands, from settling in the Big Woods. He was a great believer in postcard power, since one showing the big trees first brought him to the Northwest. He secured packets of postcards showing Pittsburgh, Pennsylvania, under a heavy smoke pall. These were mailed to eastern friends who were displaying entirely too much interest in beaver country, identifying the view as his town of Portland, Oregon.

Filling eastern speaking engagements, the lumberjack historian talked about the constant rain that grew webs between the toes and caused feet to mildew. He warned that the Indians might attack at any moment (many Easterners at the time still believed that the Indians remained a threat in this outlying wilderness) and that the Hanford atomic plant was causing the sterility of most every male in the region.

"Kept a lotta people from moving out here," Holbrook said in grim satisfaction.

Asked about the great Vanport flood of 1948, which made world headlines by wiping a wartime city of government housing off the map, Stewart shrugged, "They have them every year; this time it just happened to make the papers."

Portland, his adopted hometown, was described as "the most backward city on the Pacific Coast, with a lack of originality that stops you dead." Portlanders were aghast, but brushed it off with, "You all know Mr. Holbrook. He doesn't really mean it." Yet they weren't so sure when he campaigned and submitted a bill to change the city's name to Multnomah, for a local Indian tribe. The uproar shook the peaks and made national news wires.

When Holbrook died in 1964, his Blaine Association seemingly went with him. But then in the concern over floods of newcomers, with terms like "environment," "livability," "quality of life," and "preservation of Oregon's beauty spots" heard on all sides, the Blaine Association was revived effectively by Oregonians as a grassroots method of combating such growth. It remained much as Holbrook intended, an individual effort without benefit of organization, dues, officers, charter, or regular meetings, but instead each person "doing his own thing" to curtail wild expansion and industrial development. Even Governor Tom McCall, whose office was by nature expected to support without question Oregon industrial expansion and a mounting tourism, took up the Blaine banner whenever he spoke before conventions and other visiting groups.

"Come visit us all you like," was McCall's war cry, "but don't you *dare* come here to live."

The towering McCall, who is six feet, four inches tall, also warned industries considering locating there that they must expect to conform to Oregon, that the state wouldn't yield to them, and that the protection of his state's coastline, rivers, estuaries, mountains, forests, desert, water, and air were of prime importance to all Oregonians. Provincialism? Yes—but a provincialism with the highest motivation. McCall's family, like Holbrook's, were New England Yankees, and the two men had many parallels. A lifelong love for his state, extending from deep roots, found McCall spearheading and endorsing many of Oregon's pioneering environmental actions of the late 1960s and early 1970s, including legislation to save the beaches and to curtail the mounting litter problem. This was much to the consternation of business interests and developers who longed for the day when McCall would be out of office, hoping that then the dams might be opened and the flood let in. His Blaine-like approach even dared stand against the powerful timber and wood products industry by warning companies dumping waste material into the rivers and air that they'd better clean up or they'd be shut down.

Governor McCall didn't believe he could halt industrial and population growth, any more than did Stewart Holbrook. What McCall and his supporters were attempting was to slow things down, to gain the needed time to have the legislature adopt the necessary protective laws to keep Oregon from ruination. And at the same time, the Oregon attitude as reflected in the Blaine movement inspired other states and localities to take a look at themselves and what a future of all-out growth promotion held for them.

The Blaine campaign was exceedingly effective, gaining much national publicity. And as the environmental movement gathered momentum, the timber industry realized that public pressure would likely bring it to task. Where once Northwesterners feared the industry because of its great dominating power over everything from jobs to legislative action, this new generation couldn't be easily pressured into abandoning its courses of action. But tough, old-line timber people, who don't like being told what to do, especially within the confines of their own forests and production plants, remain unwilling to admit that the changes that have come to logging and wood processing are results of environmental pressure.

"The need was there, we had the know-how, and the time was right," one timberman tried to explain. But would the industry have executed a widespread cleaning up and improvement of its methods of doing things without the heavy pressure? He refused to answer that; however, it is doubtful, even were it economically profitable, as was the salvaging of the Tillamook Burn of World War II, that the timber industry would have moved on its own volition to improve its logging and manufacturing practices without this public pressure. Would the Tillamook Burn have ever been salvaged if it hadn't been for the wartime log prices of that time? For at least a decade, lumbermen spokesmen have been announcing that they were phasing out the wigwam waste burners, but just how long would it take without environmental pressure and clean-air laws? Some 10 years ago, when I was working on a sponsored history of the old West Coast Lumbermen's Association, the idea of having a symbolic scene of an old sawmill with wigwam burner on the cover was rejected "because we are phasing them out." But a decade later, that phaseout is still going on. As in helping Nature grow trees much faster, the constant prodding by the environmentalists has helped speed the industry and public servants in doing a far better job in the forests, reducing waste to a bare minimum, and getting the maximum from every tree, even use of the shade by hikers and other outdoorsmen who are allowed to enjoy the benefits of the private tree farms more or less as they do the public lands.

Full utilization . . . multiple-use forests . . . a better environment through

forestry . . . forests for the future . . . high-yield forestry . . . the environmental impact . . . environmental quality . . . a wide array of new terms
and phrases describe and designate the rapidly changing course of affairs in the
Big Woods. At the peak of the 1971 Oregon environmental action, the legislature adopted a tough new Forest Practices Act; the first in the nation, it
is since being used as a pattern for other states, including Oregon's northern
and southern neighbors. It replaced the 1941 conservation act, rated at that
time as very advanced legislation in support of the timber industry to offset
prospects of pending federal regulation of the forests. The new Forest Practices
Act is geared to the 1970s, not only insuring continued wood productivity but
also facing up to public concerns over air and water quality, wild and aquatic
life, and aesthetics. It is a hard law, aimed at the entire scope of forest
cutting and reforestation, born on the high water mark of the 1971 session
when so much was being passed to protect the state's environment, and then
signed significantly that summer by Governor McCall at the dedication of the
new Western Forestry Center. Rather than dwelling on specific practices, the
law sets minimum standards for better logging, stream care, road construction
and maintenance, harvesting, application of chemicals, and the disposal of
slash. Moreover, the act embraces private as well as public lands, and it fits
well with the work of other agencies directly concerned with resources and the
environment, and in achieving the maximum benefit, for every segment of the
population. Under the act, strong measures can be taken to enforce its provisions; when a violation is discovered, the state forester can call a halt to any
further destruction and make the operator repair the damage.

"It's the toughest, most advanced act of its kind in the nation," observed
one industrial forester. "It will improve regeneration, for if anybody still has
cut-and-get-out ideas, he's going to be fined so heavily it won't pay him to
do it."

Under the fire of outspoken public demands for cleaner rivers and air, the
timber industry has spent many millions on air, water, and land pollution
reduction. Oregon laws covering these areas are on a par with the Forest
Practices Act and are described as "equally tough." Not all the timber outfits
accepted these new laws gracefully; there was much breast-beating, as there often
is with other industrial groups when faced with laws forcing major changes
in their methods of operation. Timber officials declared that it was impossible
to meet the standards at all or within the allotted time; that they would have
to shut down, throwing many people out of work; and that the huge expense
would break them. A few openly defied the new laws, like a Portland roofing
company, long designated among the worst air-pollution offenders. Authorities
not only closed down the plant but also threatened to take the midwestern
owner to court and even put him in prison if the firm didn't comply—which
the company did finally, after exhausting every conceivable delaying tactic.
There was some static from workers being laid off, but public officials stood
their ground and—something the company didn't count on—workers like the
loggers have many sportsmen within their ranks whose sympathies are strongly
with the protection of the Oregon outdoors, even when it might cost them
paychecks, for many thousands are living in Oregon to be close to the forests
and the sky, and accept lower annual wages just to do so.

Most of the wood products industries recognized that it's a far different world from yesteryear and that the old-time sawmill and woodworking plants, like the loggers of two decades ago, were fast becoming legend. There are so many college graduates in logging and forestry running around that old-timers find it difficult to be understood. And the environmentalists and preservationists were snooping everywhere, aggressively examining everything, including the log booms in the rivers, which have long been one of the familiar sites of the Big Woods.

Oregonians were engaged in a massive cleanup of the Willamette River, which flows through the heartland of the sprawling Willamette Valley. With a burgeoning population in the valley, wood manufacturing and agricultural processing plants became prime targets. The vision was, and is, development of a so-called Willamette Greenway, to be preserved, like the Oregon coast, for public recreation on the river and along its banks. This wasn't possible if the river remained a sewage ditch. The processing plants would have to change their ways. The public and pollution authorities also went after farm field burning, which each summer created billowing clouds of choking smoke across the valley, south of Portland to Eugene, 125 miles distant, causing crises in centers like Eugene, where doctors advised people with respiratory problems to get out of town. Indeed, the smoke thrown up by the field burning seemed far worse than that from logging slash burnings of a generation before.

Yet each new demand, each tightening of the laws, each new "minimum standard" inspired ingenuity somewhere, for these lumber and logging people have always done things this way. Requirements for cleaner air and water posed particular problems for kraft mills and plywood plants, especially with their tangy odor, which screwed up noses of travelers through those areas. Oregonians were becoming so odor-conscious that they even attacked the delightful aroma of a coffee plant, forcing it to install pollution-control devices. The kraft and plywood problem spawned an inventive new company called Wasteco at the tiny hamlet of Tualatin. Wasteco began dealing with hard-headed practicality in the matters of incineration and sewage treatment, and in transforming waste into something useful, which would thus eliminate the dumpage into rivers and the air.

The matter was becoming serious. Antipollution codes were threatening the plants with closures and shutdowns; in one typical case, U. S. Plywood was refused permission to start another dryer at its Willamina plant, reducing the work force by some 40 men. At Albany, recognizing the necessity to keep operating, Boise Cascade went along with Wasteco, installing a system that would burn sander dust for dryer heat, eliminating dryer exhaust and disposal of the dust. The innovative and sophisticated incinerator proved to be an in-dustrial pacemaker, although when an environmental agent first talked to the nonsmoking plant manager about pollution, the manager found it particularly irritating that all the while, the agent was filling the office with choking smoke from his cigarettes. No matter; in the end, the plant met standards better than did the environmental agent.

The new unit stood an impressive 23 feet high and 10 feet in diameter, its great steel tubes bringing 1,500 pounds of sander dust per hour to the cylindrical combustion chamber. With automatic heat and flow controls, the incinerator utilizes a new suspension burning principle, where the waste is fed at high velocity and burned at an extremely high temperature. Furthermore, the new system brought about savings in natural-gas consumption, an important item in the energy crisis, with the Albany plant becoming the first in the nation and perhaps the world to utilize all its sander dust from its veneer drying, thus cutting costs on both dryer fuel and waste disposal. It won plaudits and citations of environmentalists and worked so efficiently that Boise Cascade installed similar equipment at other plants in what became the latest advancement in the manufacture of plywood, developed within 60 miles of where the product was born.

More than any other group, save perhaps the oil and automobile industries, the timber industry finds itself wide open to attack by environmentalists and preservationists. The industry is still living down its lurid past and therefore is viewed as untrustworthy. No matter what is done, it isn't right; there are ulterior motives in any movement, no matter how philanthropical it might appear. So allege the environmentalists, who simply don't trust the timbermen— and that includes most everyone connected with the industry. Even the magnificent new Western Forestry Center has been charged with being a timber propaganda tool existing on taxpayers' land. Unlike Europe, where forestry is an honored profession, Uncle Sam's state and federal foresters are believed the servants of the industry, particularly the big private companies, which control millions of acres in the Pacific Northwest alone. The companies are found, in the eyes of the preservationists, to be greedy, selfish, and bent on ruining the forests, the environment, and the country to pile up huge fortunes for themselves and their stockholders. The companies use their great power to shape legislatures, influence Congress, control state boards of forestry and the U. S. Forest Service, and to apply heavy pressure to break down barriers to the harvesting of additional timber on public lands—overcut, say the environmentalists. That the yearly logging limit is established by law has no bearing; the timber barons will find a way around it, allege the critics. Then the timbermen will cut and get out, for they are no better than their forebears.

Actually, the story in the 1970s isn't much different from other segments of American life, where suddenly and with a frightening finality, many chickens are coming home to roost. Great sweeps of prime agricultural land have been turned into housing developments, forcing states to pass laws preserving farmland for the growing of crops. In like manner, thousands of acres of national forest land were being set aside as wilderness and recreational preserves beyond the reach of the power saw, at a time when the situation was so tight that mills of the Big Woods were shutting down for lack of logs and homebuilders were scratching for studding.

Timber has been cut in the national forests for many decades with scant public concern. During the Great Depression, no such logging took place, since it would compete with private timber operations. Following World War II, tracts of national forests were harvested on a bid basis of direct stumpage sales to private operators. That's how they are always handled, since the federal government isn't directly involved in logging. Still, there was no public outcry; in fact, many forested areas seldom felt the trod of a human hiking boot.

Now all that has changed, for the new generations not only love the outdoors but also are using it. The available forest land for harvesting is shrinking, therefore, by its conversion to other uses and needs. Furthermore, the new viewpoint has also brought a complete about-face in priorities and values as to what is really important for the America of the future. For most of two centuries, everything has had an economic price tag. But now the emphasis has shifted to the environment on the social values of forests, rivers, lakes, deserts, and seashores. The aesthetic worth has always had a high price tag in the Pacific Northwest; now the need seems stronger than ever to preserve the grand views of unblemished forests, rivers, and snow peaks, to lift the spirit of the noise- and smogbound city dweller, even though he may seldom venture very far into the back country. Just knowing that a clean, unblemished wilderness is there or talking to someone who has "been" seems to help his morale, as does lifting his eyes to the jagged peaks of the Cascades from his high-rise office building. Being prevented from seeing such scenes by smog angers the Northwesterner who has long had other reasons beyond turning a buck for living where he does.

In that other world of lumbering, people of the Oregon Country have put up with a lot, long tolerating not only the unbridled forest fires and slash burnings, but also broadside clear-cutting, which ruined streams and eroded the slopes. They may not have liked it, but it was a way of life amid times of plenty, and an economic necessity. It is still the latter, for timber remains the leading industry of Oregon, employing 85,000 people, while the state of Washington relies on other leaders, timber being fourth in line. But the approach is different now; the industry itself, along with state, federal, and private forestry experts, has shown that it can survive and supply the nation's wood needs and those of part of the world through a far better way of doing things. Still, it is difficult, timbermen find, to get their message across on forests for the future. Americans have been lied to so much in recent years by their governments, public leaders, and business and industrial giants that it becomes difficult to accept honest information even when spread before them in plain language.

"I can't stand what's happening to our scenery," declared a retired college professor who fancies himself a full-fledged ecologist and has no love or trust of the lumber industry. "It makes me mad."

He was thinking of clear-cutting, the long-established system of harvesting in the Pacific Northwest that became the subject of a raging debate in 1972–73 when a western congressman opened up that particular can of Joe Cox wood worms on the national scene. To the public mind, particularly that of the layman who has scant understanding of logging and forestry techniques, clear-cutting is hideous, wasteful destruction like war bombing of a portion of our national heritage accomplished by uneducated highbinder roughnecks who care for nothing except transforming these wonderful trees into fortunes. "An unsightly mess" is the way a layman looks at it, even if he has grown up in the Big Woods. Even the wife of a leading forester expresses her distaste of clear-cutting—and she should know better. Part of this attitude goes back directly to an inbred dislike of seeing a tree felled—*any tree*—because it is a thing of graceful beauty, which by its longevity becomes something of a permanent landmark, closer to God than most of the lesser plants and creatures, including man. The poetic words of Joyce Kilmer have had a deep effect on many Americans.

"I just don't like to see a tree cut down because it's a living thing," remarked Larry Williams, executive secretary of the Oregon Environmental Council, which continually needles and challenges the timbermen and the foresters. Williams expresses popular sentiment.

On the other side of the coin, the head of one of the nation's leading timber corporations, based in the Pacific Northwest, declared:

"What we really need to do now is something about Joyce Kilmer. . . ."

The impact of that touching poem and lovely song that "only God can make a tree" is much more devastating than many professional timber people like to admit. It is almost a fighting slogan of the environmental conflict of the seventies, like "the Battle Hymn of the Republic."

"One thing I can never understand is the attitude people have toward cutting trees," says Bill Hoskins, reforestation forester for the new Tillamook Burn forest. "Trees are to be cut and used, and regrown again."

U. S. Senator Gale McGee, a Wyoming Democrat, touched off the wild, angry debate over clear-cutting upon viewing the barren messy slopes of his own state and neighboring Montana, where lodgepole and ponderosa pine and Englemann spruce had been stripped from hillsides, which were slashed with zigzagged bulldozer roads. Although after some years an effort to reforest had been made, the southern slopes were especially damaged from eroding soil, the hot sun of summer, and the brutally cold winds of winter, so that few seedlings could survive. McGee called it a "rape of the land," a crime of violence in which the Forest Service was guilty. His remarks seemed vaguely familiar, echoing those of Franklin Roosevelt decades ago at Grays Harbor. McGee proposed a national 10-year moratorium on clear-cutting as a timber harvest practice for public lands, and introduced legislation in Congress that would ban the method throughout the country.

The action shook the Douglas fir belt, since clear-cutting has long been an accepted practice as the best possible method for the rapid harvesting and speedy regeneration of timber in this rain-soaked region, where trees grow like spring weeds. The Douglas fir needs sunlight to grow, along with cool, damp air. In sharp contrast, clear-cutting doesn't work well in the pine country east of the Cascades, nor in southern Oregon, where the summer sun is too hot, burning the newly planted seedlings. But Northwest foresters look upon clear-cutting as a "silvicultural tool" in the Big Woods, consistent with modern forestry's approach to multiple use of the public lands and quick regeneration of harvested sections as protection against soil erosion and the growth of other tree varieties of little or no value. Logging operations today must include well-rounded impact studies and reforestation plans before a tree is cut. The quicker a Douglas fir can take root, the faster the forest will return once again to enhance the region as the timber grows to another harvesting in 40 to 60 years. This is an important part of modern forest management.

The idea of outlawing clear-cutting, which also has its economic advantages for the industry, threatened the Northwest timber belt with turmoil. Environmentalists charged that the timbermen wanted free use of this method only to reduce costs, since the harvest could be done as swiftly as cutting a grain field with today's power saws and efficient yarding and loading equipment.

Careful logging of individual trees, long called "selective cutting," took up far more time and was more trouble. In such critical situations, I have long felt that the timber companies are their own worst enemies, especially when they clear-cut in full public view along busy highways. Old Alex Polson sensed this a generation ago when he planted flowers to dress up the unsightly scenes around his logging operations. The ancient image of cut-out-and-get-out and damned-if-we-care crops up again. When clear-cutting was done a few years ago along Oregon's busy Sunset Highway to the beaches, it was a case in point. In one tract, a corridor of standing trees was left to shield the angry scar from public view. Then the highway people came along and realigned the route, wiping out the tree shield. Horrified by the sight, people wrote to the newspapers and public officials and environmental leaders, attacking the timber owner—indeed, all loggers and lumbermen—and placing their sympathies with ecologists who advocated more restrictions. But the damage was already done. Up the road apiece, an independent operator logged his tract right to the same highway, ignoring pleas from state foresters by roaring emphatically, as only a rugged logger individualist can express it, that they were his trees and he didn't give a damn what the public thought about it.

The U. S. Forest Service has been severely criticized for its clear-cutting on public lands, often because people motoring through the mountains hate to look across a scene of thick timber and then run up against a clear-cut, appearing like an unsightly scab on an otherwise gentle view. Recognizing the value of aesthetics, federal foresters concluded that if the clear-cuts were made less unsightly, public opposition might be moderated. They discovered that if, instead of straight-line block patches, the clear-cuts were blended into the contours of the hillsides and mountains like natural clearings, public static was reduced measurably. Now the Forest Service uses landscape architects to plan its clear-cuts.

"I wish, too," said Robert H. Torheim, U. S. Forest Service deputy regional forester, "that people would get out of their cars and walk into the clear-cut areas to see what is happening—the growing new forest."

Many foresters, both public and private, yearn for the same consideration, since they spend much time protecting their right and left flanks against amateur environmentalists who make charges without getting all the facts. In one instance, environmental leaders merely flew over a clear-cut site and then made charges in the newspapers that the logging operation was ruining a key watershed area. However, when guided over the logged-off section where the tiny seedlings could be seen taking hold, their case fell apart.

But private foresters and representatives of the timber companies believe nobody has the right to tell them how to harvest their trees and manage their forests. Bill Hagenstein, the executive vice president of the Industrial Forestry Association and burly protégé of the late Colonel Greeley, emphatically defends this point of view: that timber is a crop and that a corn field looks pretty awful, too, "when there's nothing but stubble and stalks left." Corn is planted and harvested in one summer, while logs constitute a crop that takes half a century to ripen. Furthermore, the speed of the timetable that laboratory foresters have achieved in the regrowing of better, healthier trees makes the clear-cut all the more reasonable in the Big Woods. Timbermen also point out that Nature has always been the greatest clear-cutter through fire and wind-storms, and often Nature leaves the land in far worse shape through these disasters. But the preservationists reject this as no argument at all, since they believe that Nature can do no wrong and that what she decides, since she sits at God's right hand, will be best in the long run.

Ecologists and conservationists are viewed by Hagenstein, who was also 1973 president of Keep Oregon Green, and other timber spokesmen as overemotional and lacking in practical knowledge. The preservationists voice great concern, nevertheless, not only for the general appearance of a logged-over area, which becomes a minor point, but also for the damage to the soil, the streams, the fish, and the wildlife. The Oregon State University School of Forestry, several years ahead of the crisis, began an extensive long-range study in the Alsea watershed to learn the impact of full-scale reckless clear-cutting and patch-cutting on the land and all it contained. Ecologists contend that the public has a right to know what the impact will be, not only on the section to be logged, but also on the surrounding area, even if the land is private, since the forest harvesting of a small acreage has its effects upon the territory for many miles. Many a war was fought in the Old West between rancher and logger over the felling of timber in a watershed that sustained the ranches in the valleys. Often preservationists feel they are being misled; that there is a coverup like Watergate of what is really happening, especially on federal lands.

"We can't get the figures; we don't know . . ." proclaims Larry Williams about the public lands. Williams once worked for the "enemy" in advertising for the big saw chain outfit Omark and is still proud of the saw chain display he developed for the Oregon Museum of Science and Industry. But, significantly, somewhere along the way, Williams swung over to the other side against the destruction wrought by the chain saw.

The same charge is laid on the private industry segment—that environmentalists really don't know what's going on there. The industry allegedly tells them only what it wants them to know, or feels that they want to hear. Some things are considered as "none of your business." Conservation enthusiasts, amateur or professional, have become major headaches for the industry; they are called "mad conservationists," "eco freaks," and "wilderness fanatics." Yet by the early seventies, it appeared the industry had accepted the fact that it would have to live with the movement, whether it liked it or not, and would be forced to spend part of those huge profits in doing a better ecological job. The key would come in making all that effort pay off.

Public education has become an important part of the picture. The industry spends great sums each year through the big companies and associations on fancy, eye-appealing brochures, newspaper and magazine advertisements, and television and motion picture documentaries to tell its story in positive ways and hopefully to win more public favor of its programs of "high-yield forestry" and "forests for the future." Like the old West Coast Lumbermen's Association operated under Colonel Greeley, the industry employs the best-available nationally known writers, artists, and photographers in highly paid public information departments to create its image on paper, film, and video-tape, for the Keep Green program decades ago proved what could be done through concentrated effort. Public education, constant and unrelenting, is a necessary thing for better understanding, even if it takes years and many fortunes. The industry is constantly embroiled in some kind of emotional public crisis, be it clear-cutting, bad logging practices, or shipping too many logs to Japan. Timber is everyone's business and at everyone's doorstep in the Big Woods. A couple of decades ago there was a screaming public debate over "selective cutting," an old practice with a new name, but now you hear little about it other than that this is another acceptable "tool" of timber harvesting. The waves of controversy over the Big Woods rise and fall with the seasons of events, but public mistrust, based on the excesses of the roaring past, is always there.

Still the lumbermen wheel and deal in six and seven figures; in the inflationary seventies, timber prices have skyrocketed along with wages and equipment costs from forced changes in logging operations. Douglas fir logs were going for sky-high prices, and even the little fellow could again find green gold bonanza in the hills. One couple on the Oregon coast took a federal stumpage contract for Port Orford cedar logs worth $400 apiece. They were logging on weekends and carefully covering their trail, so that pirates wouldn't move in on them.

Profit gains of the big companies for 1972 were significant for the times: International Paper, over the $2,000,000,000 mark for an industry "first," with a sales advance of $124,000,000 making it the sales leader; Georgia-Pacific, $1,920,000,000, a sales gain of $481,700,000; Weyerhaeuser, a gain of $376,200,000, to $1,670,000,000. These were steady increases over the previous years of the seventies, largely from a housing and construction boom, with the 1971 value of construction work on new private residential buildings, including farm, hitting $42,100,000,000, compared with $31,700,000,000 in 1970—a reflection of 2,051,000 units nationally in 1971, against 1,433,600 in 1970. Sales were up from 5 to 23 per cent, as were net profits, from 8.92 per cent for Georgia-Pacific to 100 per cent by Pope & Talbot, which was taking on new life in growth and development and an expanding trade abroad. In 1971 the value of cut lumber in Oregon alone was $875,000,000—7,340,000,000 board feet at an average of $117 per 1,000 board feet; and the manufacture of plywood was over 8,000,000,000 square feet at a wholesale value of $576,000,000. The bonanza of the Green Desert is still there, yielding tremendous sums as a strong indication that the Big Woods, far from depleted, is green and growing second, third, and fourth crops.

Yet these same staggering sums spent not only for research, reforestation, better equipment, and promotion, but also in construction of plush new multi-million-dollar high-rise headquarters buildings, keep environmentalists suspicious of the over-all goodwill aims and purposes of the industry. The industry, they contend, is still destructive and hasn't advanced far from the cut-out-and-get-out yesteryears. It remains robber baron in character, if not personality, is bent on exploitation, and is very much a part of the grand American tradition of turning resources into dollars for owners and stockholders, and to hell with what might happen to the country and the American people, who have a huge stake in these resources. When a timberman announces that "we'll cut all that we damned please on our own land," it makes the conservationists furious, for the timberman may also be pressuring hard to open up more federal and state lands to harvesting—a dangerous overkill in the eyes of the preservationist—trying his damndest to break loose stands within areas saved for outdoor recreation, watersheds, and scenic values. He will also campaign and push and shove against locking up valued forests as wilderness areas, now totaling millions of acres, which will remain natural and untouched so that future generations may see and enjoy the natural forest as it grew on this continent before the arrival of the white man and his destructive activities. The timberman sees no value in this, only sheer waste of a much-needed supply for the national woodpile. When one lumberman suggested they log within the boundaries of famed Crater Lake National Park, because timber in that beauty spot would die and be utterly wasted, environmentalists and conservation groups pricked up their ears, convinced that the lumbermen are panting to work the national parks and believing that all that was needed was to get a foot in the door, whereupon the national park philosophy would go down the tube.

State and federal foresters and their aides find themselves caught in the center of these continual controversies, and in many instances, federal foresters especially are indicted right alongside the timbermen, the popular belief being that they are puppets, fully controlled by the industry. No matter what action the industry takes, it is wrong according to some segment of the environmental movement. When the Sierra Club reported that Weyerhaeuser was planning to phase out its huge Pacific Northwest holdings in favor of those abroad, in what was charged as the old cut-out-and-get-out style, company president George Weyerhaeuser denied vigorously that any such plan was even considered. Yet when I suggested to an environmental leader who had been extremely critical of Weyerhaeuser's operations, that maybe this would be a good thing, since perhaps the lands would be managed more to his liking by a new owner, he grimaced and retorted hoarsely, "I would call it irresponsible on their part."

The environmental war, from timber's view, is a case of being damned if you do and damned if you don't. Timbermen would like environmentalists and preservationists to view things in better focus, as to what the country is and what its demands are each day of every year. They charge, and with justification, that many so-called environmentalists and their outspoken supporters often don't understand, and don't see the broad picture. A major part of their complaint is that environmentalists aren't objective and aren't certain of their information and therefore jump to wrong conclusions; or because they aren't neutral, they reach the conclusion first and then gather only evidence in its support. For instance, a leading preservationist alleged that the private tree farms of one big company weren't open for public recreation. Not so, replied company spokesmen in a simple telephone call. They have developed picnic, camping, and recreational places and built trails for public use; the woods are wide open save during critical fire weather or prior to the Yuletide, when too many overly ambitious folks might set themselves up in the Christmas tree business, at the owner's expense.

"One thing that disturbs me," declared Don Finney, resource manager for Ketchikan Pulp Company, in a debate with environmentalists, "is that we get opposition from the Sierra Club people . . . on practically everything we do. If we want to go into an area and put a log dump, they just categorically come out and oppose us. It costs us thousands of dollars to go through all the motions and all the agencies to overcome the objections to a log dump. Eventually we do, because the objections are unfounded . . . but what's happening is that we're spending money . . . that comes out of the stumpage value, because that's the way stumpage is figured. It's an operational cost. When you do those things, you're forcing us give up some money that we could use for environmental things. What I'm saying is that there is an obligation on these people's part to know what they're objecting to."

Finney's position finds much support within the Big Woods, where lumbermen describe environmentalists as "six hanging from every rock" of the Pacific Northwest. Eager student-research teams and aggressive individuals reach instant answers after interviewing the wrong people, from inexperience or trying to build a certain case. It doesn't take much to gain a headline or television time for someone with something on his mind, off-base as it might be. Youth seeks instant results, especially in this pushbutton age, failing to consider that it takes, even now, 40 to 60 years to regrow a forest. It has taken a quarter century for the Tillamook Burn to become green and beautiful again. At the same time, open debate and the people's right to know are integral parts of the American system so that the ecologists and preservationists play a necessary role as watchdogs over the Big Woods. Yet lumbermen and career foresters are also concerned about this same future and have been doing something about forest preservation. The advances in forestry and lumbering since the destructive times are unbelievable, and especially within the past two decades. Things being accomplished now as everyday fare are far beyond the extended vision of the most thoroughly oiled and imaginative lumberjacks logging half a century ago in Erickson's Saloon or Aberdeen's Hume. The tree farm concept, which at the time critics branded as a "publicity stunt," has grown into a vast, valuable, and highly workable tool toward the goal of perpetual forests, not only in growing timber but also in sheltering wildlife, protecting watersheds, preventing soil erosion, and as a treasured source of oxygen for the increasingly poisoned air.

Still, it is strange that in the midst of timber optimism for the future, a critical lumber shortage existed in the early seventies, brought about by intensive twin housing booms in the United States and Japan. Moreover, many Northwest forests are in a state of regrowth, not yet quite ready for harvesting, as are many public forest areas. Lumber was indeed in short supply in the spring of 1973, with prices skyrocketing and lumberjacks resorting to costly logging methods, such as back-country helicopter lofting to meet the demand.

"I can't get any two-by-fours," moaned one dealer, in a surrounding of trees. "It's virtually impossible, and the price is terrible. I'm movin' my office and was going to sell an old desk. I had one corner propped up with a piece of two-by-four. Told a friend I wanted to sell the desk. 'To hell with the desk,' he replied. 'Sell the two-by-four.' "

I commented that I had several in my basement.

"Keep 'em," he advised. "They may become collectors' items."

Sawmills were shutting down from a lack of logs, while the timber companies were being blamed for selling sticks by the millions to Japan, where they could get three times the value of finished lumber on the U.S. market. Between 2,500,000,000 and 3,000,000,000 board feet were shipped in 1972, and in the state of Washington, $38,000,000 was generated through sale of state timber to the Japanese. And the Japanese were out shopping in the Big Woods, not merely for logs but also for sawmills, among them one at Tillamook on the edge of The Burn.

The clear-cutting hassle seemed forgotten—for a time, anyway—as industry, the public, and environmentalists became caught up in the issue of logs for the Japanese. Much of it was tied to U.S. foreign policy and the declining value of the dollar abroad and the unsettled balance of payments; but the impact in the Big Woods went beyond timber, the lumber shortage, and the skyrocketing cost of housing in the pressures being brought upon existing forests. Weyerhaeuser announced that it would expand its 1973 harvest to meet the demand; other interests were accused of trying to pry open the federal forests and appeared to have succeeded when the Nixon administration increased the annual cut by a billion board feet, purportedly to ease the lumber shortage for housing and overseas balance of payments. This is the kind of danger conservationists fear, when the pressures and political hustling become sufficiently heavy to unlock forests too soon and too much. At the same time, there was pressure on the President and Congress to outlaw log shipments to Japan, the action backed by workers' unions and construction companies, but opposed vigorously by the longshoremen, who wanted the work. The feeling was that the home country needed first to be served, and legislation was introduced by Bob Packwood, Oregon's junior senator, to halt the shipments. Then the Japanese volunteered to curtail their orders, and in the shock of Watergate, the issue seemed to subside.

The management of Pacific Northwest forests is a seesaw game, as complicated as a moon shot. The industry is sensitive, packed with emotionalism, and as reactionary as Wall Street. Conservationists help keep the balance as custodians of the woods, and therefore serve a most important function. By the same token, wood is a basic need for survival, and therefore extremists who propose locking up all the forests, never cutting another tree, are as off-base in their utopian dream as is the timberman who would like to fall every available tree. The environmentalists are constantly reminding the public that the forests must be preserved; that historically, every great civilization that has depleted its forests has gone into a tailspin, hell-bent for oblivion. Yet there is a difference, for the United States of the twentieth century isn't wholly dependent upon wood. Its basic energy has come from oil, the fossil fuel, while other important resources have been the metals of the ground and power from the rivers. The industries involved with these resources, until lately, have come under less criticism than the timber industry, save when there is a gigantic oil spill, as at Santa Barbara, or when mining operations gouge out beauty areas or threaten scenic places. Yet when the oil is gone and the last ton of mineral is extracted from the ground, it will all be *gone*—completely used up. Americans were reminded suddenly of this reality by the gasoline shortage.

In contrast, timber is the only natural resource that can be regenerated at the present time, anyway, and that is what is happening. Some 75 per cent of the nation's forest land at the time of Columbus is still in trees—about 759,000,000 acres, a third of which is set aside in parks, forest preserves, and watersheds. Over 10,000,000 acres are classified as wilderness and primitive areas. In 1970 private landowners planted 1,206,700 acres of future growth; and the bulk of the timber land is not held by huge, powerful timber companies, but is in the hands of small, private owners, such as those with farm woodlots.

While we fiddle, Rome seems to burn in the energy crisis; the curtailment and rationing of gasoline will likely have a long, heavy impact on the easy freedom Americans have enjoyed, where they could leap into their cars and take off for long distances to the mountains, the beaches, Las Vegas, or San Francisco. This country's most serious shortages are in other resources, not in the growing of timber. By the year 2000 the nation will need double the wood products it now uses. But by then, what about energy? Will it be coal . . . or solar energy . . . or will it come from the sea? In the long run, the use of natural gas, gasoline, and oil seem all to be on the wane. There was a time not so long ago when the common carriers of the Big Woods were powered by coal and cordwood, and thereby steam, on the rivers and along the iron trails, and sawdust waste heated the homes of the Pacific Northwest and other timber areas. Wood use for fuel and power may indeed come back, for the chances are that we will need to rely on a wide variety of energy sources to survive, rather than putting all our chips on a single square.

Several years ago, my good mother-in-law, a pioneer-type lady raised during central Oregon's raw days just after the turn of the century, installed an oil circulating heater in her farmhouse, hopefully to make things slightly easier in her later years. But I had the feeling she never quite trusted the darned thing, for she didn't get rid of her wood circulator nor her wood cookstove. When threatened by a winter oil shortage, she stoked her woodshed solid with chunks, as she had done in years past. She was determined to be warm and comfortable. And she may yet come up winners.

The trees of the Big Woods are out there, green and growing, thicker, healthier, stouter, and more safely contained than ever before. The tools are at hand to keep them growing. It's now a matter of management . . .

Chapter XVII

Epilogue:
Seventy-three Million Trees

It's beautiful, isn't it? . . .

J. Edward Schroeder
Oregon state forester
to the author,
October 1972

In midautumn not long before this book was completed, I traveled by bouncing four-wheel-drive truck with forester Bill Hoskins to a high ridge of the Tillamook Burn country, where you could look for many miles across some of the most ruggedly beautiful country on the American continent.

The land below unfolded in blue-green haze into craggy ridges and deep canyons and sharp peaks clear to the Oregon coast near the town of Tillamook, rising and falling in mighty convulsions that had taken place many centuries ago. Wispy clouds of white fog were drifting from the canyons like smoke of long-dormant fires that had perhaps been smoldering for decades. Yet that rising moisture moved lazily upward until it disappeared into the blue haze, giving off a serenity to the place, below a bright yet thinly overcast sky.

But the sheer beauty of the setting wasn't this singular view. It was the trees. Close at hand and far below, stretching for miles upon miles and climbing en masse to the high pinnacles of this amazing land, was a young yet already grand new forest. I hadn't been in the heart of the Tillamook Burn for a number of years, and the change was startling. Wherever your eye scanned the region, in all directions, healthy, growing trees had replaced what had been utter desolation.

"This winter [1972–73] we will complete the planting under the bond program," Hoskins said casually. "By then we will have planted 73,000,000 trees."

Think of it—73,000,000 trees! And that didn't count the 73,200 pounds of treated Douglas fir seed scattered by helicopters throughout this 300,000-acre site of one of the greatest forest fires of them all, a territory of ash and rubble, which many who were wise in such matters said was impossible to restore. Yet here it was, the job gloriously completed; and within five years, Hoskins added, they would begin harvesting the first crop of Douglas fir, which was now around 12 inches in diameter.

For years, like thousands of others, I had driven through The Burn (they call it the Tillamook Forest now, although it will always be The Burn to a lot of us) along the main highways to the Oregon coast. Until very lately the scene had been much the same—the black snags and stumps and fallen deadfalls, the rubble of downed logs, the brush and alder, and the bright color of the vine maple in its autumn dress. Only lately had you noticed that evergreenery was crowding out the brown and yellow of the deciduous trees along the highways and on the immediate slopes. Yet I had the feeling that in the interest of good public relations, the highway areas had been planted first so that the scars of the great fires could be hidden from public view. My mind was still back there in the past, fixed on snags and deadfalls and logging trucks and dust. I was, therefore, totally unprepared for the sight below.

We had come into The Burn through the old logging railroad stop of Owl Camp, renamed for Nels Rogers, toward this viewpoint, which had served as a heliport during 'copter seeding of this particular area. My mind went back to that day in 1949 when a small crowd of us—foresters, lumbermen, loggers, newsmen, the curious—gathered in the open air of Rogers Camp to launch the huge replanting project—and to even earlier years of reporting the last two big fires; the discouraging hearings, which debated what to do with this sorry region, leading to the initial experimental projects in snag falling, road and fire trail construction, plantation clearing, hand planting, and aerial seeding. Often I was squired about by Rudy Kallander, the forest rehabilitation director, along steep, narrow, back-country logging and forestry roads to watch fallers and buckers working with cumbersome power saws or to see brawny men stooping and bending, their gnarled hands placing tender shoots of young seedlings into doubtful soil, where the shoots were only faintly visible to the trained eye amid brown, bracken fern and low brush cover. When a section was finished, it appeared no different than it had been for at least two decades.

I had seen the first broadcast seedings by helicopter and use of the whirlybird to build a strategic new lookout to protect the new forest from fire. And I watched thousands of schoolchildren, Scout troops, and church groups scrambling from squatty yellow buses to spend a day in specially designated tracts "helping to reforest the Tillamook Burn." The public became involved, and thousands of Oregonians feel that in some small way, they helped to bring back this forest.

Seventy-three million trees . . . when it first began, and these good foresters told me what they planned to do with all that mess, I went away shaking my head. Like many others, I found the whole idea overwhelming. The region was too huge, too wild, too ravaged by constant fire. At best, it was a gamble, and no one now alive would ever see its resolution.

Yet year after year, without fanfare, they went on planting trees, struggling along the steep, windy slopes and braving the chilling ravines. Tract after tract, in shadow and sunshine, in rain and in snow, the seedlings went into the soil, from the Gales Creek canyon to the South Fork Camp and down on Jordan Creek and along the high ridges near the old Reeher's Camp. The Tillamook Burn reforestation became largely forgotten to the general public, for there is little drama in the planting of trees compared to big-city riots and the killing of a President. The world and American thinking have changed much since Governor Douglas McKay stood on that old stump at the Owl Camp in 1949 to dedicate the project. When it began, Harry Truman was President. We were recovering from World War II, with veterans crowding the college campuses, and living under the shadow of possible atomic war with the Soviet Union. Jet planes, television, and freeways were novelties, virtually unknown in the Pacific Northwest. There was a fearful new war in Korea, while in The Burn, small crews were setting trees for a future that seemed far away. Through the turmoil of the past few decades, through the upheaval and convulsions this country has experienced, the Northwest has been steadfastly loyal to its own values. We landed on the moon, not once but several times, and in a far corner of Oregon, unheroically, they went right on planting trees. Now it is done, and it is a happy sight indeed.

Seventy-three million trees . . . 226,000 acres have been stocked by hand and from the air—109,000 acres in hand planting and 117,000 acres by helicopter. The crews even scaled the steepest slopes, their corks clinging precariously to the hillsides. But the evidence was there; the trees were showing color. Most every part of the main Burn was reached in some manner, save for a few isolated places that were considered impossible or an occasional chunk of land missed by the chopper pilots.

"For a long while," explained Hoskins, "we couldn't figure out why areas like that one over there beyond this heliport didn't sprout trees. Then we discovered the reason. The pilots weren't dropping seed there, probably because of the way they took off. Once those boys were in the air, we had no control over them."

Other spotty acres will regrow from natural seeding as present trees mature and begin to reproduce their own kind. Despite all that happened, this mist-washed region is still foremost for growing timber. One place, along Cook Creek and the East Fork of the Trask River, was giving planters trouble. It was so hot there during the 1933 fire that nothing seems to grow, but foresters are still trying to bring it back.

Within a few years the distant vistas will no longer be easily seen, for the thick trees are beginning to block the views. Driving along the wheeling roller-coaster roads, you can still see signs of the fires in charred stumps, an occasional bleached snag, and the black remains of old logging railroad trestles. But signs of the tragedy are fast fading now as the new forest reaches for the sky.

Many of the changes aren't so obvious. The streams are clearing, for the new forest is holding back the soil and also the moisture, making The Burn far safer from fire. Yet it remains fire-prone to an extent, as is any forest. The district has been plagued by small fires along the old PR&N railroad to Tillamook; and in 1968, 40 acres were burned along one ridge from man's carelessness. That had potential of another big one, a year short of the "jinx year," but forestry crews hit it hard, got it out, and then lost no time in replanting the scarred slope.

Salvage logging has long since been completed inside The Burn and has shifted to tracts owned by the state on fringes that escaped the fires. These tracts are quickly replanted. But before too many years, logging will return to The Burn, a fact that shocks some people. However, this is part of the over-all plan. Thirty-five per cent of the timber sales is retained by the state to support The Burn, while the rest goes to the counties. In the future, areas will be logged each year and then replanted as part of the multiple-use system for this great new American forest. The hope is that The Burn can "go it alone," and it was significant that in the 1973 state legislative session, the Forestry Department, unlike most public agencies, asked for no more money to support the project. But the funds must come from somewhere for administration, road maintenance, fire protection, recreational development, and all the rest. And foresters aren't missing a trick at putting the new forest to work without damaging the over-all endeavor. Two Bonneville Power transmission rights-of-way cut through the center of The Burn, bound for coastal cities. You can't allow full-size timber to develop beneath high transmission lines, so the foresters grow Christmas trees, leasing out the plots to people in the business.

If all goes well, the Northwest District should have little difficulty meeting future budgets, and neither should Tillamook County, where the bulk of The Burn is located. On the investment in rehabilitation funds, a return of $30,000,-000 is expected in revenue from the original forest crop. In 1970 Tillamook County received $533,483 as its share of the annual income of $845,457. By the year 2000, The Burn's projected annual income will reach $1,257,171, of which Tillamook County will receive $924,021; and 100 years hence, the projected income could reach $4,532,891, with $3,576,451 going to the county.

"Tillamook County stands to become the wealthiest county in Oregon," Hoskins said. "I don't know what they'll do with all the money."

The new Tillamook Burn Forest, dedicated in the summer of 1973 by Governor Tom McCall, is for far more than commercial harvesting. Recreation is now a big item, where it wasn't considered in the discussions of the 1940s. Because of its huge size and close proximity to one of the West's major population areas, the demands of the future are expected to be heavy. The Burn is already being heavily used and the back roads traveled by hikers, cyclists, and horseback riding groups. Where The Burn appeared once as foreboding to anyone but the determined hunter and outdoorsman, its roads busy with frightening log trucks, its sheer beauty is now again an open invitation to the public. And the widespread and varied use of this new forest of the Big Woods has the long-range attention of state forester Ed Schroeder, his assistant Frank Sargent, and their aides; with Schroeder and Sargent and the others who worked on the project in its hopeless days, the salvation of The Burn remains a personal achievement.

"It is seldom in a man's lifetime that he is able to see something like this happen," Schroeder told me. "It makes a fellow feel pretty proud to have been a part of it."

A lot of Oregonians, including the thousands who expressed faith with their votes that it could be done, or who planted a few seedlings as schoolchildren, feel the same way.

The Burn, which claimed the lives of millions of wild animals, birds, and fish during the big fires, is once again an important wildlife habitat, not only for deer but also for black bear, coyote, beaver, mink, the smaller woods animals, and many varieties of birds, including the increasingly rare bald eagle. Herds of stately Roosevelt elk roam the forest in increasing numbers. Salmon, steelhead, and trout have returned to the rivers and streams; the Wilson River has long been considered one of the finest steelhead rivers of the Northwest for this seagoing rainbow trout. There are unusual geological formations that attract rock hounds searching for jasper, agate, quartz, and crystals. Sandstone beds yield specimens of leaf prints and seashells, for all this was once ocean bottom. And there are legends of lost gold strikes and buried treasure from Spanish galleons, which crashed centuries ago on the rocky coastal shores to the west. In spring and summer, wild flowers bloom profusely along the slopes and gullies, which were once only black, unforgiving ruins; while in the autumn, the hills still turn the bright, gay colors of the hardwood trees against the background of the rising new evergreens. Foresters will be happy when the new forest takes over with a vengeance, for one of the big problems of babying along The Burn is keeping back the rapidly growing alder and other brush forms that attempt to crowd out the conifers.

Multiple use—those are the key words for the future of this rejuvenated, parklike forest, as it is for most every woodland of today. By 1973, 10 parks and 300 campsites had been developed, plus boat ramps, waysides, back-country trails and viewpoints where people on foot, horseback, motorbike, or in cars may view some of the Northwest's most awesome scenery and the varied wildlife. There are plans, too, for a museum of artifacts, pictures, films, tape recordings, and other materials to preserve the full story of the Tillamook Burn for the appreciation of future generations—for once the trees have become large and tall, the scars will fade from view, and the record of this tragedy and its amazing comeback after long struggle will be in danger of becoming lost, outside the public libraries.

"We want people to discover The Burn and explore and enjoy it," remarked Bill Hoskins. "That's part of what it is for."

When we headed toward the highway, I glanced over my shoulder through the rear window, back at those wonderful evergreens scaling the hillsides. As Ed Schroeder said, *it is beautiful*. This is where it all began, where forestry and logging came together and turned around, entering a new, enlightened age. This was the front-runner. . . .

Suddenly, I realized I was looking at this grand scene across Bill Hoskins' corks, riding high and proud behind the front seat. Logger's boots . . .

How was it Nelson Reed put it in his well-taken *Ode?*

> "Such men as they
>   have made this country great. . . .
>   Pray God, Oh Stranger, others
>   yet be born worthy as they to wear a logger's boots."

Logger's boots . . .
And beyond, 73,000,000 trees . . .

Bibliography

Books:

Abdill, George. *This Was Railroading*. Seattle, Wash.: Superior Publishing Company, 1958.

Adams, Kramer. *Logging Railroads of the West*. Seattle, Wash.: Superior Publishing Company, 1961.

——. *The Redwoods*. New York: Popular Library, 1969.

Allen, Alice Benson. *Simon Benson: Northwest Lumber King*. Portland, Oreg.: Binfords & Mort, 1971.

Andrews, Ralph W. *This Was Logging*. Seattle, Wash.: Superior Publishing Company, 1954.

——. *This Was Sawmilling*. Seattle, Wash.: Superior Publishing Company, 1957.

——. *Timber: Toil and Trouble in the Big Woods*. Seattle, Wash.: Superior Publishing Company, 1968.

Binns, Archie. *Sea in the Forest*. Garden City, N.Y.: Doubleday & Company, 1953.

——. *The Roaring Land*. New York: Robert M. McBride & Company, 1942.

——. *The Timber Beast* (a novel). New York: Charles Scribner's Sons, 1944.

Breetveld, Jim. *Treasure of the Timberland*. New York: *Scholastic* magazine, in cooperation with the Weyerhaeuser Company, 1967.

Churchill, Sam. *Big Sam*. Garden City, N.Y.: Doubleday & Company, 1965.

Clark, Donald H. *18 Men and a Horse*. Portland, Oreg.: Metropolitan Press, 1949.

Clark, Norman H. *Mill Town*. Seattle, Wash.: University of Washington Press, 1970.

Coman, Edwin T., Jr. and Gibbs, Helen M. *Time Tide and Timber*. Stanford, Calif.: Stanford University Press, 1949.

Corning, Howard McKinley. *Dictionary of Oregon History*. Portland, Oreg.: Binfords & Mort, 1955.

Cour, Robert M. *The Plywood Age*. Portland, Oreg.: Binfords & Mort, 1955.

Engstrom, Emil. *The Vanishing Logger*. New York: Vantage Press, 1956.

Fuller, George W. *A History of the Pacific Northwest*. New York: Alfred A. Knopf, 1945.

Fultz, Hollis B. *Famous Northwest Manhunts and Murder Mysteries*. Elma, Wash.: Fulco Publications, 1956.

Greeley, William B. *Forests and Men*. Garden City, N.Y.: Doubleday & Company, 1956.

Guthrie, John A., and Armstrong, George R. *Western Forest Industry*. Baltimore, Md.: The Johns Hopkins University Press, 1961.

Hays, Finley. *Crown Zellerbach Loggers*. Portland, Oreg.: Crown Zellerbach, 1970.

Hidy, Ralph W.; Hill, Frank Ernest; and Nevins, Allan. *Timber and Men: The Weyerhaeuser Story*. New York: The Macmillan Company, 1963.

Holbrook, Stewart H. *Burning an Empire: The Story of American Forest Fires*. New York: The Macmillan Company, 1945.

———. *Dreamers of the American Dream*. Garden City, N.Y.: Doubleday & Company, 1957.

———. *Green Commonwealth*. Seattle and Shelton, Wash.: Simpson Logging Company, Dogwood Press, 1945.

———. *Half Century in the Timber*. Seattle and Aberdeen, Wash.: Schafer Logging Company, Dogwood Press, 1945.

———. *Holy Old Mackinaw*. New York: The Macmillan Company, 1938; rev. ed., 1956.

———. *Tall Timber*. New York: The Macmillan Company, 1941.

———. with Jones, Nard, and Haig-Brown, Roderick. *The Pacific Northwest*. Garden City, N.Y.: Doubleday & Company, 1963.

Horn, Stanley F. *This Fascinating Lumber Business*. Indianapolis, Ind.: The Bobbs-Merrill Co., 1943.

Hosmer, Paul. *Now We're Loggin'*. Portland, Oreg.: Binfords & Mort, 1930.

Jensen, Vernon H. *Lumber and Labor: Labor in Twentieth-Century America*. New York: Farrar and Rinehart, 1945.

Johansen, Dorothy O., and Gates, Charles M. *Empire of the Columbia*. New York: Harper & Brothers, 1957.

Jones, Nard. *Evergreen Land*. New York: Dodd, Mead & Company, 1947.

Kemp, J. Larry. *Epitaph for the Giants*. Beaverton, Oreg.: The Touchstone Press, 1967.

Kirk, Ruth. *Exploring the Olympic Peninsula*. Seattle, Wash.: University of Washington Press, 1967.

Kneiss, Gilbert H. *Redwood Railways*. Berkeley, Calif.: Howell-North Books, 1956.

Kyne, Peter B. *Cappy Ricks, or the Subjugation of Matt Peasley*. New York: H. K. Fly Co., 1915.

Labbe, John T., and Goe, Vernon. *Railroads in the Woods*. Berkeley, Calif.: Howell-North Books, 1961.

Lavander, David. *Land of Giants*. Garden City, N.Y.: Doubleday & Company, 1958.

Loggers Handbook. Portland, Oreg.: Pacific Logging Congress (various editions).

Lucia, Ellis. *Head Rig*. Portland, Oreg.: Overland West Press, 1965.

———. *Tough Men, Tough Country*. Englewood Cliffs, N.J.: Prentice-Hall, 1963.

McCullough, Walter F. *Woods Words*. Portland, Oreg.: Oregon Historical Society and Champoeg Press, 1958.

Morgan, George T., Jr. *William B. Greeley: A Practical Forester*. St. Paul, Minn.: Forest History Society, 1961.

Morgan, Murray. *Skid Road: An Informal Portrait of Seattle*. New York: Viking Press, 1951.

———. *The Last Wilderness*. New York: Viking Press, 1958.

———. *The Northwest Corner*. New York: Viking Press, 1962.

Morgan, Neil. *The Pacific States*. New York: Time-Life Books, 1967.

———. *Westward Tilt: The American West Today*. New York: Random House, 1961.

O'Connor, Harvey. *Revolution in Seattle*. New York: Monthly Review Press, 1964.

Pollard, Lancaster. *History of the State of Washington*. New York: American Historical Society, 1937.

Renshaw, Patrick. *The Wobblies: the Story of Syndicalism in the United States*. Garden City, N.Y.: Doubleday & Company, 1967.

Reppeto, Paul. *Way of the Logger*. Chehalis, Wash.: *Loggers World,* 1970.

Ripley, Thomas Emerson. *Green Timber*. Palo Alto, Calif.: American West Publishing Company, 1968.

Spencer, Betty Goodwin. *The Big Blowup*. Caldwell, Ida.: The Caxton Printers, 1956.

Stegner, Wallace. *Joe Hill: The Man Who Chose to Be Shot*. Garden City, N.Y.: Doubleday & Company, 1950.

Stevens, James. *Green Power: The Story of Public Law 273*. Seattle, Wash.: Superior Publishing Company, 1958.

————. *Paul Bunyan*. New York: Alfred A. Knopf, 1925.

Teale, Edwin Way. *The Wilderness of John Muir*. Boston, Mass.: Houghton Mifflin Company, 1954.

Special Articles:

Barendregt, Peter A. "Helilogging—A New Sound in the Woods," *Loggers Handbook,* Pacific Logging Congress, 1972.

Betts, William J. "Alex Polson—From Gold Miner to Millionaire Lumberman," *West,* Oct. 1972.

"Chain Saw Story, The," *Chain Saw Age,* May 1954.

Cronnemiller, Lynn F. "Oregon's Forest Fire Tragedy," *American Forests,* Nov. 1933.

"Dialogue: Timber Industry Meets the Sierra Club," *Alaska Construction and Oil,* Feb. 1973.

Dymond, Lura. "Splinter Groups," *Westways,* Nov. 1972.

"Giant in the Chain Saw Industry, A," *Chain Saw Age,* Aug. 1965.

Hagenstein, W. D. "Emotions Aside, Clearcutting Is Silviculturally Sound Concept," *Forest Industries,* Dec. 1970.

————. "Environmentalists Add Fourth Dimension to Clearcut Harvest," *Western Conservation Journal,* May–June 1971.

Holbrook, Stewart H. "Daylight in the Swamp," *American Heritage,* Oct. 1958.

———— with A. (Whiz) Whisnant. "The First Fifty Years," *Loggers Handbook,* Pacific Logging Congress, 1959.

"It's a Beginning," Crown Zellerbach *Times,* Centennial issue, 1970.

Lucia, Ellis. "Oregon Is 'Recarpeting' Vast Burn," *Washington Post,* Apr. 23, 1950, repr. in *Congressional Record.*

————. "Oregon: Laboratory for Forest Research," *Northwest* magazine, *The Oregonian,* Aug. 1, 1948.

————. "Top Man in Oregon's Hot Spot," *Northwest* magazine, *The Oregonian,* Nov. 16, 1952.

————. "Youth Plants the Tillamook Burn," *National Parent-Teacher,* Mar. 1954.

McKenna, Gail. "Shanghai Hell," *West,* Feb. 1972.

Miller, Charles I. "History of Chain Saws," *Southern Lumberman,* Apr. 15, 1949.

"Omark 'Epic' in Industry," *Chain Saw Age,* Jan. 1968.

Pement, Jack. "Whatever Happened to Dorothy Ann Hobson?", *Oregon Journal,* Aug. 1972.

"Power Saws Come of Age," *The Timberman,* Oct. 1954.

Richards, Leverett. "Helicopter Logging," *The Oregonian,* Nov. 1971.

Roberts, Mary L. "The Great Tillamook Forest Fire of 1933" (special reports), writing in the *Washington County News-Times* and the *Oregon Journal,* Aug. 1933.

Schroeder, J. E. "From Burn to Tillamook Forest," special report, Oregon State
    Forestry Department, rev. 1970.
Spence, Morton. "Clearcutting: Boon or Bane to U. S. Society?", *Oregon Journal*,
    Apr. 1972.
Stevens, James. "From Black Candles to Green Glory," *Columbia*, Feb. 1952.
Strite, Daniel D. "Up the Kilchis," *Oregon Historical Quarterly*, 1971–72.
Van Syckle, Ed. "Early Grays Harbor, Range of the Ramparts," *Aberdeen Daily
    World*, Jul. 31, 1963.
Woolley, Carwin A. "Bull Teams to Balloons," *Portland* magazine, Apr. 1972.

*Various issues, special articles, and reports were considered from the following pe-
riodicals and newspapers:*

Aberdeen (Wash.) *Daily World.* Special editions: March 10, 1926, and July 31,
    1963.
American Forests
Chain Saw Age
Crow's Forest Products Digest
Forest History
Forest Log, published by Oregon Forestry Department
Forest Products Journal
4-L Lumber News
Journal of Forestry
Loggers World
Lumberman, The
National Wildlife
Nature magazine
Oregon Historical Quarterly
Oregon Journal
Oregon Purchasor
Oregonian, The
Pulp and Paper
Seattle (Wash.) *Post-Intelligencer*
Seattle (Wash.) *Times*
Statesman, The (Salem, Oreg.)
Tacoma (Wash.) *News-Tribune*
Tillamook (Oreg.) *Headlight-Herald*
Timberman, The
Totem, The (Washington Department of Natural Resources)
Washington County News-Times (Forest Grove, Oreg.)
West Coast Lumberman
Western Conservation Journal
Westways

Dozens of reports, studies, pamphlets, brochures, etc., prepared by the U. S. Forest
Service, state forestry departments, timber companies and allied industries and associations, and environmental groups were considered. The author relied heavily on
his own extensive files for chapters concerning the Tillamook forest fires and the
reforestation program. Written material, notes, and tape-recorded interviews on how
the great Tillamook fire of 1933 began are retained by the author.

Index